ECO
HOMO

HOW THE

HUMAN BEING

EMERGED

FROM THE

CATACLYSMIC

HISTORY OF

THE EARTH

ECO HOMO

NOEL T. BOAZ, Ph.D.

BasicBooks
A Division of HarperCollins*Publishers*

Published by BasicBooks,
A Division of HarperCollins Publishers Inc.

Designed by Elliott Beard.

Library of Congress Cataloging-in-Publication Data

Boaz, Noel Thomas.
    Eco homo / Noel T. Boaz.
      p. cm.
    Includes bibliographical references and index.
    ISBN 0-465-01803-3
    1. Human evolution.   I. Title.
  GN281.B618   1997
  599.93'8—dc21   97-14604
                                                      CIP

97 98 99 00 01 ❖/RRD 10 9 8 7 6 5 4 3 2 1

To my mother, Elena More Anson Taylor,
a woman for all seasons

CONTENTS

ACKNOWLEDGMENTS

This book was begun while I was a Meyerhoff Visiting Professor in the Department of Environmental Sciences at the Weizmann Institute of Science in Rehovat, Israel. I acknowledge the intellectual stimulation of that environment and the immensely valuable personal discussions with Steve Weiner, Aldo Shemesh, and others that gave me a new appreciation of a broadened multidisciplinary approach to paleoenvironmental research and hominid evolution. I have also benefitted from discussions with Stanley Ambrose, Raymonde Bonnefille, Lloyd Burckle, Peter de Menocal, Greg Retallack, the late Peter Williamson, and many colleagues and former students too numerous to name. As always, I thank my wife, Meleisa McDonell, and my children, Lydia, Peter, and Alexander, for their support and forbearance.

Figure 1

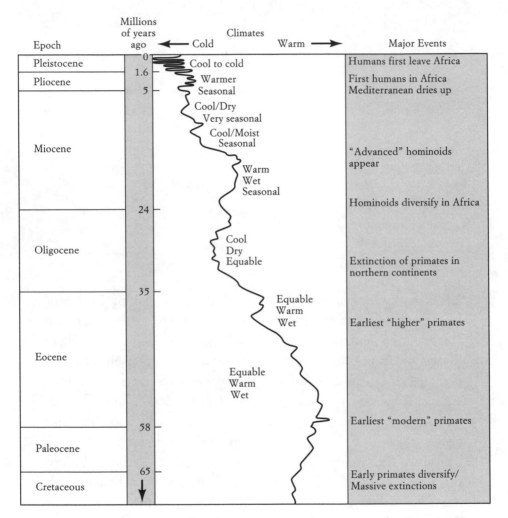

An overview of the time scale of oxygen isotopic changes over the Cenozoic Era.

INTRODUCTION

For at least the past half century, we have known one part of the story of human evolution: the part that tells us that the earliest humanlike creature—species we call australopithecines—evolved from apes, and that we, modern humans, evolved from the australopithecines, and that both evolutions occurred in Africa, the first between 4 and 6 million years ago; the second, less than half a million years ago. What remains to be answered are a series of "why?" questions.

Why, for example, did these two evolutions occur in Africa and not in Europe or Asia or South America? Why did the first evolution lead to creatures with the features of australopithecines and the second to creatures with the features we have? Why did each evolutionary event occur at the time it did and not sooner, or later?

The goal of this book is to answer some of these "why?" questions, relying on some extremely interesting scientific discoveries of the past decade. Many of these discoveries have come about as the result of the application of sophisticated high-tech tools, developed for biologists and physicists, to the study of human remains; others are a consequence of a low-tech change in how we collect fossils. But underlying both is a shift in our attention to a part of the evolutionary picture science once ignored.

To understand that change, consider the following. The prevailing visual image of human evolution is that of a plodding train of progressively more erect hominid silhouettes, an image first popularized in a Time-Life book, *Early Man,* written by F. Clark Howell and Maitland Edey in 1965. Indeed, today, this image is still so pervasive that almost any illustrated treatment of human evolution is likely to include some version of it. The figures march through time—snapshots of human existence separated from their respective ancestors and descendants by white space on the page. But, in fact, there wasn't white space between each silhouette.

We now know that very dramatic changes in the world of the early humans drove each and every evolutionary event depicted in that train of increasingly more sophisticated hominid figures, dooming some unlucky populations to extinction while allowing others, by dint of good genes, good fortune, favorable habitats, or perhaps just one good water hole at the right time, to survive, reproduce, and perhaps take their place in the evolutionary lineup. Indeed, the rapidity with which certain of the characteristics that define modern humans evolved, such as our very complex brains, attests to the terrible force with which natural selection acted on our ancestors, allowing only a few individuals, the fittest, to survive, reproduce, and evolve. Until very recently, however, we had only general ideas about what those forces of nature might have been.

Of course, our ancestors were affected not just by great—indeed cataclysmic—geological and environmental changes; they lived virtually every day of their lives very much at the mercy of nature. Even small changes in their physical conditions, we now know, the kind we experience every day without notice, could have deadly consequences.

It was not until the early 1950s, however, that anthropologists studying human fossil remains began to take an interest in the ecological world of early humans. Before that, fossil hunters took lit-tle or no notice of plant or animal fossil remains at the sites they studied and made no attempt to study the geology of fossil sites for what the soil might tell them about changing ecological condi-tions. Credit for redirecting anthropological thinking to the value of this ecological perspective goes to two anthropologists, whose work we must turn to next, if we are to understand how later sci-entists began to unravel the "whys" of human evolution.

A REVOLUTION IN ANTHROPOLOGY

In 1953, ecologist George Bartholomew and anthropologist Joseph Birdsell, both of UCLA, wrote a paper entitled "Paleoecol-ogy and the Protohominids" that created the new field of paleo-ecology. In that paper, they set out many of the basic assumptions and generalizations that would guide the investigation into the en-vironments and adaptations of our early forebears to this day. Most important were their thoughts about the limited abilities of early man to control his environment. Bartholomew and Birdsell suggested that our earliest ancestors, much like every other small to medium-sized mammal on the African savanna 4 to 6 million years ago, spent most of their time just trying to stay alive—striv-ing to eat and to avoid being eaten, responding to local changes in water and food, searching out safe sleeping places for the night, mating, and caring for their young. Therefore, to understand hu-mans and the life they led, we must learn everything possible about their physical world.

In the forty years since the original paper was published, Bartholomew and Birdsell's original ideas have remained remark-ably up-to-date, though, of course, later scientists have built and expanded upon this foundation. One of the most important ways they did so was in developing the multidisciplinary paleoanthro-pological research project to test out the hypothesis of early man as a hostage to his environment.

Development of the
Multidisciplinary Research Expedition

Although Louis Leakey developed over many years an impressive team of specialists working on his site of Olduvai Gorge, Tanzania, it is F. Clark Howell who first pulled together the modern paleoanthropological research expedition. Here, a coordinated team of scientists examines a potential fossil site, each member focusing on one independent yet interrelated point of reference. While such an approach can be expensive, logistically challenging, and difficult to manage (the French archaeologist François Bordes laughed once that all those people dressed in khakis in the field looked to him like a lavish hunting safari), it often results in highly effective exploration of a site.

My own exposure to Howell's multidisciplinary approach took place in graduate school at Berkeley in the early 1970s. In 1973, as part of a team Howell put together, I traveled to southern Ethiopia to do field work. At first, I must admit, I was less than delighted with his approach. Digging in dirt for 2-million-year-old fossils sounds better in the telling than in the doing, especially under Howell, who seemed to have an insatiable interest in what can only be described as the minutiae of the fossil record. For sure the number of fossil teeth that my excavation team was uncovering daily began to mount—from the dozens to the hundreds—but none of these fossils came from human remains and I saw myself as a human fossil hunter. I became frustrated by the dawning realization that a new skull of *Zinjanthropus* (the Leakeys' "Dear Boy" discovered in 1959) was never going to turn up in my excavation. I even began to question why I was at the site at all. This was backbreaking and boring work in the heat of the African summer and all we were finding were isolated bits of animals. Who cared, I thought. I actually remember telling my excavation crew chief, a Kamba farmer named John Kaumbulu, that I would rather be across the border at Richard Leakey's site, where they were finding complete skulls of hominids.

Howell, however, seemed to revel in the flood of teeth coming from the excavation. He sensed that I was less than thrilled with how the work was going and one day as we were cataloguing the haul, he uncharacteristically chastised me for failing to recognize the tooth of a new fossil equid (a zebra) that we had discovered.

But then the day arrived when the first hominid came out of the excavation. I heard the muttered word *mtu* ("man" in Swahili) as several of the excavators gathered around a bit of fossil tooth enamel. At first it looked no different than all the other bits and pieces of antelopes, hippos, monkeys, and other animals we were finding. No glowing radiance set it apart. But then a jolt of excited recognition struck me as I looked at the tooth and realized that the thick enamel cap and low rounded crown was that of a hominid. Maybe the increased adrenaline got my neuronal synapses working or maybe I had already partly figured out why Clark's approach was the right one, but since that day I began looking upon hominids not as something apart but as just one of the other animals in the site. The excavation had spoken to me and its message had been that hominids are animals—intrinsically linked to the other denizens of that ancient African landscape of 2 million years ago, all linked to each other and their common environment. Once I understood that, I also understood why a multidisciplinary approach is so important.

What, you may ask, do the other members of such a multidisciplinary team do? The geological team will determine the exact layering of the rocks in which the fossils that you investigate are found. This work is called stratigraphy. Every single fossil that is discovered must be relatable to this stratigraphic scheme so that its age relative to fossils from other stratigraphic levels can be accurately assessed. The geological team also is responsible for collecting samples of rocks and minerals for later laboratory absolute dating (determining the age in years), for collecting samples for investigating the chemistry of the ancient soils and sediments, and for undertaking any other earth-science studies that may be relevant.

Once a stratigraphic framework has been established, the paleontological survey team, composed of individuals who are trained observers and who can be relied on to religiously and accurately record every collected specimen, goes to work. It is not always easy to see fossil bones in the broken, partially covered, and disarticulated states in which they are exposed by erosion.

A specialized paleontological team is the microfaunal team. Microfauna are remains of small animals that even in the gentlest of fossilization conditions are discovered only as individual teeth. This team collects sediment from a fossil locality that is suspected to be productive of these tiny fossils. This sediment is then put onto fine-mesh screens and washed with a jet of water to remove all the silt and clay. What remains after washing and drying is then laboriously sorted by hand and microscope to find the small teeth and bones. Small animals provide an immense amount of data on the paleoecology of a site because they live in restricted habitats with clearly demarcated environmental limits. Their population numbers can also be sensitive indicators of environmental change.

Rounding out the field team are specialists on fossil plants, sometimes pollen experts, or palynologists, and experts in fossil wood, or paleoxylologists. Pollen and fossil wood found in sediments can tell much about the past environment since in most cases they are the same species alive today. Like the microfauna, their numbers vary through time in response to ecological changes.

Collecting the data is only the first step. Graduate students, funding organizations, and sometimes even the scientists themselves tend to forget that after the field data are collected—the dramatic part—they must be analyzed. This usually takes years. As aspects of anthropological research have become more "high tech," a much higher proportion of the research is done in the laboratory. Isotope studies of sediments and fossils, for example, help establish absolute dating. A scanning electron microscope can study microdamage on fossils—it has a hypothetical finger that literally runs along the surface of the fossil, recording every indentation, yielding information about where the fossil was originally deposited and how much it has moved since deposition. Sometimes bones are purposely broken or damaged in the laboratory to match the state of preservation seen in actual fossils. Sometimes these studies lead to derivative or secondary fieldwork, so-called "actualistic studies," in which processes at work today are studied in known ecological contexts in order to see how today's potential fossils reflect current conditions.

One might think that as anthropology has become more laboratory oriented the field component of research would become deemphasized, but in fact the opposite has occurred. The old research maxim of "garbage in, garbage out" is never more apt than when sophisticated, high-tech methods of analyses are used on improperly collected, poorly controlled, or inadequately prepared specimens. If anything, there is now a greater premium placed on rigor in fieldwork.

ISOTOPES AND ENVIRONMENTAL CHANGE

At the same time that the anthropologists were getting serious about collecting ecological information at fossil sites, paleontologists Stephen Jay Gould and Niles Eldredge, with common origins at Columbia University in New York, promulgated a new view of evolution that would lead some earth scientists to look for dramatic ecological disturbances that might be correlated with major evolutionary changes. This theory suggested that evolution was characterized by long periods of little or no change, termed *stasis,* and periods of rapid change, termed *punctuational events.* These authors claimed that hominid evolution was a prime example of this type of punctuational evolutionary change. Puncutated change came about through dramatic environmental disturbances.

The first proof that a cataclysmic geological event may have had a dramatic evolutionary response came when evolutionary geophysicists Walter and Luis Alvarez at Berkeley Lawrence Laboratory found a chemical signal indicating that a meteorite had hit the earth just at the time the dinosaurs disappeared, suggesting that it had wiped out the dinosaurs in one fell swoop. It would be years before other scientists finally provided confirming evidence that such a sequence of events did in fact occur, and still today there are holdouts.

Generally speaking, a problem that all attempts to demonstrate a catastrophic mode of evolutionary change face is that there are many gaps in the fossil record as well as in the record of environmental change. Are the significant changes in a species lineage from one time to another an artifact—a result of a big gap in the fossil record—or is the change in reality a relatively sudden event? Critical to this research effort then is finding a continuous record of change. These sorts of data have been very difficult to obtain from terrestrial geological deposits, simply because earth movements and erosion over time have played havoc with the processes of long, steady laying down of sediments by bodies of water or other processes of sedimentation. Fortunately, earth scientists, looking to establish a better and more continuous climate record, had decided as far back as 1967 to drill down deep into the earth's crust in the open oceans around the world and to bring up to the surface for analysis a series of long tubes of sediment. This project was called the Deep Sea Drilling Project. It has been indispensable in the effort to obtain a well-dated and continuous record of climatic change.

In 1955, Cesare Emiliani at the University of Chicago studied drilled columns of seafloor sediment—actually he studied the shells of four separate species of the tiny ameboid sea animals known as Foraminifera found in these columns of sediment—and reported that two isotopes* of oxygen—^{16}O and ^{18}O—varied in

---

*Isotopes (Greek for "same place") are variants of a single element, which always has the same number of protons in its nucleus, but a variable number of neutrons. Some isotopes, such as carbon 14 (^{14}C) and potassium 40 (^{40}K), are unstable and undergo a standard rate of radioactive decay, useful in dating. Other isotopes are stable isotopes, and these exist in equilibrium in nature. It is these latter isotopes that are most useful in studies of paleoclimatology.

their proportions by identical amounts throughout the length of a core. Through some very interesting work, he was able to establish that whenever there were high $^{18}O/^{16}O$ ratios, one knew that glaciers had been spreading on land and global temperatures had been lower. The opposite trend was seen when warm, interglacial conditions returned.

Using Emiliani's pioneering effort and the tools of isotopic geochemistry, other researchers have extended the oxygen isotope record back through the entire Cenozoic era, the last 70 million years or so. This record is of the utmost importance in beginning to understand climatic change in the geologic past because for the first time it provides us with quantitative measures of global ice volume, translatable into past global temperatures. The use of oxygen isotopes has now become commonplace in researching climatic change, and many nuances of the technique have been developed. Other elements are now also used and these can provide checks of the oxygen isotopic record or provide totally new insights on past climate, as we shall see.

Isotopes are now also being used to investigate localized climate changes. The deep-sea core is best relied on to reveal global patterns. Local effects get masked because the world oceans are constantly being mixed and homogenized by moving water currents. Air movements on land, however, may be much more restricted, leading to quite localized climatic regimes on land. Thus, for example, the uplift of a mountain chain might block rain-carrying winds, leaving an area on the leeward side of these mountains arid, regardless of the patterns of global climatic change. Isotopic records preserved in sediments in this area can be used to investigate whatever the local climate was like. These uses are of particular interest to anthropologists because it is climatic change affecting local areas of the Old World tropics and then successively higher latitudes of continental landmasses that is crucial in understanding the ecological history of humanity.

Thus, there is a grand attempt afoot to understand at a much more fine-grained level past environments, the history of climatic change, and the mechanisms of ecological change both globally and within different regions of the world. Many scientists, from climatic modelers using supercomputers to pollen experts dredging up cores from African lake beds, are working on these problems. It is an exciting and fast-moving field of research and will occupy in various forms much of the rest of this book.

Defining Communities of the Past

Regardless of how detailed and accurate our paleoclimatic data are, they tell us little about a past species' adaptations and evolution unless we can confidently place that species within the environmental context to which the data pertain. J. Arnold Shotwell was a paleontologist at the University of Oregon who pioneered taphonomic approaches to relating animal fossils to particular habitats within ancient animal communities. Shotwell's excavation methods and his analytical technique allowed him to separate those animals whose bodies had become buried close to where they had lived and animals whose bodies had been transported into the fossil deposit from somewhere else.

First, Shotwell distinguished between communities of living animals, termed *biocenoses,* and communities of dead animals, burdened with the name of *thanatocenoses.* This differentiation of where animals had lived from where they died was a fine point glossed over by earlier generations of paleontologists. Of course, their environmental reconstructions were also very coarse grained—if a fossil fish had died, was deposited, and subsequently found in a Carboniferous coal deposit, it was obvious that it had *lived* in the precursor to that deposit, a freshwater swamp. Similarly, if one discovered a mastodon in wind-blown glacial sediments, then it had lived there in the glacial environment. What was the big deal with taphonomy?

The big deal came in with an increasing appreciation by paleontologists of ecology—the interaction of animal species with each other and with the environment. Yes, the fish in the coal deposit had probably lived there, but so had the ancient amphibian found in the same deposit. Yet the amphibian had legs, and it probably lived in the water only part of the time, leaving the swamp to venture out on land. Thus, the amphibian in life had been a member of a watery community and a terrestrial community (two biocenoses) but its body had ended up in one thanatocenosis with the fish.

Shotwell wanted to figure out how to tell whether particular animals had come from the same biocenosis when they were found in a fossil deposit. He started from general principles. If a perfectly complete skeleton of an animal with all bones intact, articulated, and undisturbed was found in a deposit, then it was very likely that the animal died in or very close to the particular environment represented by the fossil deposit. On the other hand, if the skeleton was very fragmentary—with bones missing, broken up by exposure to the elements, chewed on by scavenging animals, or washed away by flowing water, for example—then probabilities were that that animal had lived farther away from the site of deposition.

Shotwell devised a simple index to test this idea in his excavations. For each animal species, he counted the number of individual bodies represented in a fossil deposit and then he counted the number of bones still present in each of the bodies. He then calculated an index of relative completeness. The species with the highest indices, and therefore the most complete skeletons in the deposit, were actual members of the biocenosis in which they were deposited. The less complete and more fragmentary skeletons belonged to animals that had been transported into the deposit from a more distant habitat. Shotwell termed these two broadly defined communities the *proximal community,* for those species that had lived close to the depositional site in life, and the *distal community,* for those species that had lived far away.

Shotwell's insight provided for the first time a measure, independent of the animal's anatomical adaptations, of where within an ancient environmental mosaic an extinct animal had lived. Anatomy still played a big part, however. In an African fossil site, for example, hippos and Nile perch may have been members of the proximal community in a particular fossil deposit, but they certainly did not have the same adaptations and they certainly utilized their shared portions of the environment in very different ways. For this type of deduction, analysis of anatomical specializations is still very important.

A question that paleoanthropologists are frequently asked is: How do we know where to look for hominid fossils? Where did the hominids live? Using a combination of Shotwell's methods and an appreciation of hominid adaptations as preserved in their anatomy, we can be confident that the earliest hominids lived in savannas, woodlands, and even forests closely tied to water. Biogeographically, the earliest hominids were restricted to Africa. Sites where sediments were deposited in low-lying freshwater basins or in caves in Africa during the Pliocene epoch are thus likely sites for finding hominid fossils. Later, the hominids spread out of Africa. By piecing together both global and regional paleoenvironmental patterns with anthropological knowledge of hominid evolution, it will be possible in the following chapters to construct a number of models of how and why hominid species evolved through time.

HOMINID ORIGINS WITHIN THE GRAND SWEEP OF PRIMATE EVOLUTION

The study of human origins must be comparative. We, as anthropologists, are already at a significant disadvantage when studying our own species because it is easy to fall into the trap of thinking we already know the answers. Humans, at least from our perspective, are unique, but if we adopt an idiographic approach—one that describes only a singular event, human origins—we have no scientific basis for testing our hypotheses or coming up with new ones.

Humans are the end product of a last-gasp evolutionary paroxysm of a dying breed, the apes, that somehow succeeded. But all our kindred apes, the gorillas and chimps of Africa and the gibbons and orangutans of Asia, are relics of an evolutionary array of species much more diverse, widespread, and successful. Primate evolution provides many important lessons for human evolution simply because it spans much more time, records many mores twists and turns of the earth's paleoecological history, and shows what sorts of species–environmental interactions took place over this period of time. It thus is instructive for us who wish to understand past human evolutionary responses to environmental change and who have so little comparative data today.

Eco Homo begins with the appearance of higher primates in the fossil record. In actuality, our evolution goes back to the beginnings of life itself, some 2 billion years ago, and for the bulk of our evolutionary history the narrative is the same for us, kangaroos, fruit flies, and slime mold. I concentrate here, however, on those parts of the evolutionary record, only the last 40 million years or so, that produced higher primates. Within that period, our purview will become progressively broader and clearer as we approach the more data-rich periods of more recent geological time. The earliest hominids are now known at dates between 4 and 5 million years ago, and the earliest *Homo sapiens* are only about half a million years old.

Because primates in general and hominids in particular are tropical animals, our narrative will also deal primarily with the environmental history of the Old World tropics—the rain forests, woodlands, and savannas that straddle the equator and adjoining latitudes. After early primates die out in the Northern Hemisphere early on in our narrative, the forbidding world of the high latitudes becomes colonized relatively late in human history. Earlier generations of theorists on human evolution—not coincidentally living almost exclusively in Europe and North America—thought that humanity had been crafted and formed by the challenging and hostile environments of the Northern Hemisphere Ice Age. But we now know that the high latitudes were not the sources of most of the evolutionary novelty in the human story. Conceived of as glacial ice sheets and subzero temperatures during which our ancestors lived in caves, the late Pleistocene Glacial Maximum is too late in time and too limited in geographical area to have had a major effect on the main course of human evolution. Rather, as Herodotus said in the fifth century B.C., "*ex Africa semper aliquid novi,*" or "there is always something new coming out of Africa."

The fossil record from the African tropics produces the earliest dates for virtually all the major stages of human evolution. Despite Darwin's remarkably prescient deduction in 1873 that earliest human origins lay in equatorial Africa, the continuing evolutionary emergence of new hominids from Africa over the entire span of the Pliocene and Pleistocene epochs—the last 5 million years—is surprising. Yet Africa has persistent and powerful ways of reminding us of our ancient connections with her—from our unabating interest in our closest living animal relatives, the chimpanzee, bonobo, and gorilla, to the fear generated by the AIDS (acquired immune deficiency syndrome) and Ebola viruses, organisms coevolved with our genome in the primordial forests of our origins. Even the visions of terrible human tragedy with which many people in the West associate Africa—the periodic famines brought on by drought and the wars that result in thousands killed—are reminders that this is our heritage. Like the magnificent faunas of the African savanna, which continue to function under the old laws of population ecology far away from the industrialized, polluted, and domesticated world of the West, the human populations of Africa live closer to the land. Natural selection, using such tools as environmental change and overpopulation, has winnowed over and over again this array of human diversity. Africa is our ancestral homeland, and even today it still contains a stunning three-fifths to four-fifths of all human genetic diversity.

Eco Homo will delve into the reasons that the hominids first diverged, a story that transpired within Africa. I have been particularly concerned with this problem over the past twenty years and I have undertaken numerous field and laboratory investigations aimed at finding answers. My answers are presented as a number of hypotheses, the best syntheses that I can suggest at the moment, and ones that will hopefully be refined or even falsified as we gather more data in the future. Hypothesis 1 (Chapter 2) concerns the split of our lineage from that of our largest cousin, the gorilla. Although we have some large fossil apes spread about in Africa over the last 18 million years or so, it is unlikely that any of them represent direct ancestors of the modern-day gorilla. They are either too primitive in their anatomy or are in the wrong place at the wrong time to be credible ancestors. Without fossil bones of direct gorilla ancestors, I construct my hypothesis on gorilla origins from the fossil evidence about their environments, the molecular evolution of the gorilla, and the anatomy, adaptations, and geography of living gorillas. I believe that the gorilla makes most ecological sense as a barrel-chested, high-altitude giant adapted to montane habitats isolated as the climate became drier and cooler. Its habitats in Central Africa were differentiated as geological doming and uplift took place in pre–Rift Valley Central Africa some 10 million years ago.

The evolutionary split of the gorilla left a creature in our lineage that we still have not found in the fossil record—a common ancestor of humans and chimpanzees. I hypothesize that this creature was a lowland species, a small-bodied form adapted to forests and woodlands, and able to range into habitats less densely treed than its gorilla relatives (and probably late-surviving apes now extinct and unknown). Recent fossil discoveries in Africa at sites older than 4 million years have begun to give us some idea of what this forest and forest-fringe creature may have looked like. Undoubtedly it was small, long-armed, with teeth that looked very much like those of the living pygmy chimpanzee. Its brain was no larger compared to its body size than that of living apes. Thus, at this very early stage we must look to other reasons than the inordinate ballooning of the human brain as the evolutionary prime mover of differentiation of the evolutionary line leading to humans. What ecological changes then could have effected the evolutionary split of this population of chimp-people into the lineages leading respectively to chimpanzees and to ourselves?

I investigate two possible hypothetical scenarios for the chimp–hominid divergence. Both hypotheses are based on geographical isolation of populations of protochimps and protohominids, and at this stage of our knowledge, both are about equally plausible. Hypothesis 2 (Chapter 3) posits that the colossal cleavage of the earth's crust known as the Great Rift Valley, located in Central and East Africa, was the mechanism separating plant and animal communities. In this hypothesis, termed by French paleoanthropologist Yves Coppens "The East Side Story," the chimpanzees evolve on the western side of the rift and the hominids diverge on the eastern side of the rift. While eminently plausible and fully concordant with the available fossil data, there may be another explanation.

Hypothesis 3 (Chapter 4) puts forward the idea that we may be seeing only part of the story by focusing entirely on eastern and southern Africa. First, eastern Africa is a relatively small part of the continent, forming a part of the narrowing that characterizes its southern half, and a bit of the northeast rim that extends up into Ethiopia. The larger part of the continent lies to the north and west—an extensive region larger than Europe whose fossil record until recently has remained virtually unknown. Second, although the African Great Rift Valley can be seen from the space shuttle, the most striking geographic feature of Africa from space is the much more immense Sahara Desert, the largest on earth. Instead, the chimp-people, living in forests and forest fringes to the north of the Central African forests, may have differentiated into the hominid and chimpanzee lines by the geographical isolating effects of the encroaching Sahara Desert. The drying up of the Mediterranean Sea between 5 and 6 million years ago, the so-called Messinian Event, would have been the prime mover of this environmental and geographical change. New fossil discoveries of a 3.5-million-year-old hominid in Chad, and my own continuing research in 5- to 7-million-year-old deposits in Libya, have begun to provide some significant support for considering this hypothesis in a positive light.

Whatever caused the initial separation of the hominid and chimp lineages between 5 and 7 million years ago, the ensuing 2.5 to 4.5 million years saw the differentiation of a successful group of hominids—the australopithecines. These hominids were bipeds, perhaps the first in our lineage to adopt this habitual upright posture and means of moving about on the ground. We know this by fragmentary but telltale bones of their extremities found in the fossil record. Although the evidence points to the earliest of these hominids retaining close ties to their ancestral forest and forest-fringe habitats, Hypothesis 4 (Chapter 5) relates the major evolutionary changes in the australopithecines to their adaptation to a progressively fluctuating climate. Increasingly longer periods of aridity and consequent spread of grassland savannas in Africa changed the australopithecines' habitat and put a premium on populations that could adapt to these changes. The most important of the anatomical changes that we see in the fossils were in the feet and legs, used to walk to more widely dispersed fruit trees and other food sources, and teeth, with flatter and heavier grinding surfaces to eat the tougher savanna foods once they got there.

In our incorrigibly anthropocentric view of the world, it may seem inevitable that the australopithecines only existed to lead to us. But these African hominoids were successful species in their own right and they lived as long as, or longer than, our own genus *Homo* has been in existence. If there was no inherent drive toward humanity, and there is certainly no reason to believe in such a thing, what then did propel the australopithecines down their evolutionary path to humans? Hypothesis 5 (Chapter 6) relates the evolution of the first humans to the beginning of increasingly severe climatic fluctuations in Africa, associated with and possibly brought on by the spread of Northern Hemisphere ice sheets. Now it was no longer sufficient to be able to simply walk from one patch of woodland to another to get food, because there just were too few of these patches around. Australopithecine populations either adapted or died out. One evolutionary solution was the genus *Homo,* a clever opportunist with an expanding brain who used tools to increase the range of savanna foods it could eat. The other evolutionary response was the robust australopithecines, hominids with huge molar teeth and massive chewing muscles that could crush between their jaws virtually any food they picked up. Early *Homo* and robusts lived side-by-side in an evolutionary equilibrium for well over half a million years.

A little after 2 million years ago, a date we are now becoming increasingly confident of, hominids made an exodus out of Africa for the first time. But only one species, *Homo erectus,* made the trek. The robusts stayed behind. Why *Homo erectus* expanded its range outside the continent of its origin is a more significant evolutionary anomaly than it at first appears, especially to most scientists whose more proximate origins are themselves in the Northern Hemisphere. Hypothesis 6 (Chapter 7) posits the beginning of a pattern of latitudinal habitat changes moving north and south in sequence with global patterns of cooling and drying periods—a "paleoclimatic pump"—that pushed *Homo erectus* out of Africa. Either the increasingly severe climate or competition with the larger-bodied and larger-brained *Homo erectus* finished off the robust australopithecines, and they disappear from the fossil record by 1.0 to 1.5 million years ago. Hand axes (the first complicated stone tools) and fire are associated with later *Homo erectus.* Recent work is showing that the antiquity of these two cultural hallmarks may extend back to the beginning point of the species.

Homo erectus was a species of early human that was discovered over a century ago in Indonesia. Further discoveries in China in the early part of this century yielded a view of a massively thick-skulled, craggy-browed, cannibalistic, fire-using cave dweller. Recent discoveries have now shown that the species appeared very early in Africa and soon afterward in Europe and Asia. Individuals of *Homo erectus* approached modern human stature, yet their brains were a quarter smaller on average than ours. Some of the first-discovered sites of *Homo erectus,* in Indonesia and China, are now apparently quite late in time, and surprisingly they are of the same age and even later in time than sites in Africa that have given up the remains of *Homo sapiens.* Did these two species of humans coexist for a time, and did *Homo sapiens* have anything to do with the extinction of its thick-skulled cousin? Ecological principles suggest that the answer is yes.

We know too little of Europe during the period of transition of *Homo erectus* to *Homo sapiens* to construct a detailed scenario for this transition, but data are much more abundant for a later period of human evolution. The Neandertal period may shed significant light on what may have transpired several times throughout Pleistocene human evolution. The Neandertals, those cave people whose popular image is synonymous with dull-witted brutishness, are now known to have been a population of *Homo sapiens* that lived in Europe and the Middle East, and one which adapted successfully to extreme conditions of glacial cold. New dates also show that Neandertal populations moved into regions peripheral to their core ranges during periods of global cooling, displacing anatomically modern *Homo sapiens* as they did so.

It appears that Neandertals won a few rounds before eventually being knocked out of the evolutionary ring by anatomically modern *Homo sapiens,* our direct ancestors. Perhaps our modern prejudices against Neandertals are the last reverberations of this ancient intraspecific conflict within our species. For whatever reason, but likely because of ecological competition with anatomically modern *Homo sapiens,* the last Neandertals were gone from Europe by 32,000 years ago. The ecological correlates of their displacement by anatomically modern *Homo sapiens* has much to tell us about previous waves of human population movements and the dynamics of human evolution in Eurasia over the last million years. Hypothesis 7 (Chapter 8) lays out my ecological argument for the extinction of the Neandertals and why the "Out of Africa" model for the appearance of modern humanity is the most likely explanation of the facts.

Culture, with a capital *C,* has always vexed scientists who study human evolution. What exactly is this system of learned behavior that allows modern societies to conquer virtually any environmental obstacle and occupy almost every corner of the earth's surface? It has been described as "superorganic," that is, above the laws of nature, but if that is so, where did it come from? Anthropologists of a generation ago believed that culture appeared at a "critical point," before which we were "animals" and after which we were "human." Both archaeology and field studies of the nonhuman primates have now shown that this conception is wrong. There was a long, slow period of development of strongly patterned learned behavior that did not vary significantly over tens and even hundreds of thousands of years.

The great florescence of material culture in the form of stone tools, ornaments, cave and wall art, and musical instruments did not occur until late in human evolutionary history. Importantly, their appearance does not correspond in either Africa or Eurasia with the first appearance of those humans we identify from their bones as anatomically modern *Homo sapiens.* Rather, it is the onset of the Glacial Maximum, which reached its peak some 18,000 years ago, that is associated with the great development of human culture as an adaptaive mechanism *par excellence.*

It is the thesis of Hypothesis 8 (Chapter 9) that ecological change, characterized by increasingly severe environmental fluctuations of greater and greater amplitude during the late Pleistocene, forced the evolution of culture as we know it. What evolved as a superior energy-extracting system for human groups in the marginal habitats of the Pleistocene was a machine that overproduced in times of plenty. Culture in the hands of the first agriculturalists 10,000 years ago increased population densities, created diseases unknown to earlier hominids, built the first villages and temples, and gave humans the pervasive misconception that they were above the laws of nature even as they rushed head-long into a despoilment of their habitat unknown in any other species. Culture evolved as a means of rapid response to environmental change—one much more rapid than the genetic response to natural selection. It saved the human species from extinction, at least in the Northern Hemipshere, when many other large mammal species died out during the late Pleistocene Glacial Maximum. Culture then is the quintessential positive feedback system, feeding on change to generate change. It is a powerful force and one that can be decoupled, albeit temporarily, from the laws of nature and natural selection.

We are now living in a warm and rainy interglacial period, but it is certain that the next glaciation will come, covering most of the northern half of North America and Eurasia with a mile-thick layer of creeping ice. The next glacial period is not going to happen for another 75,000 years, however, and it is consequently not going to make urban planners and futurists lose any sleep. Futurists, who are particularly concerned with designing a sustainable ecological future for *Homo sapiens,* have a problem with predicting trends past the year 2010, when straight-line predictions of current trends become fuzzy. Can the long-range perspective of *Eco Homo* help them in confronting such problems as global warming, pollution, deforestation, the loss of biodiversity, and indeed our ecological survival?

1

Ecological Changes and Primate Evolution: The Prime Movers of Change

The long search for the ultimate causes of human emergence from the natural world has been a tortuous one. It has been difficult for self-important human beings to accept that they originated as small, inconspicuous animals in the shadow of the dinosaurs.

Contemplating the scene from my window looking onto the Oregon High Desert, it is easy to imagine this ancient time. I can see many small mammals much like our distant ancestors scurrying about in the juniper scrub and up in the trees. The large grazing ungulates like bison never made it this far west, and the scene that one sees today is a drier and colder version of the early primate North America of 50 to 60 million years ago, before the evolution of large mammalian herbivores. Evolutionary origins from little ratty tree-living animals are humble beginnings indeed for the cerebral giant primate that later came to dominate the earth. But there is an additional insult to human ego. All of the early primates that inhabited North America, our relatives and ancestors, went extinct. The primates themselves are missing from the indigenous small mammal faunas of North America and most of the rest of the Northern Hemisphere.

In the Mesozoic period, extending from some 230 to 70 million years ago, our ancestors were shrewlike small mammals that were nocturnal and tree-living. They were much like the living tree shrews of Southeast Asia but they probably lived throughout the northern half of the world (Laurasia) since climates were warm and there were no major barriers to their wide dispersal. The first primates were little-changed descendants of primitive insect-eating mammals such as these. We know about them primarily from sites in western North America. Long-snouted, whiskered, sharp-eyed, and hyperactive, these little creatures were understory denizens of worlds lorded over by huge reptiles.

When the reptilian ecological dominance of the earth ended, primates took part in the evolutionary diversification of nonreptile animals that followed. Primates, whose insectivore ancestors had lived on the forest floor among the trees, now took to the trees. The earliest primates, called plesiadapoids, evolved forearms that could be twisted around and hind limbs that could grasp, in order to climb trees and find food. A related group of archaic primates, the paromomyoids, experimented with another way of getting from tree to tree—flying. These early primates competed for air space with evolving birds, descended from the now extinct dinosaurs, and bats, their mammalian relatives. Perhaps the relative ease with which we humans learn to "fly"—airplanes, helicopters, and hang gliders—as well as our dreams of flying, a favorite of psychoanalysts, are mental attributes that we have inherited from our archaic primate relatives.

The archaic primates were very successful. At one site in western North America they comprise 39 percent of all animal fossils recovered—by far the largest single group. They dominated the forests and they proliferated into a number of different species with a wide variety of adaptations. Evolutionary biologists call this an *adaptive radiation*—an evolutionary fanning out of species from one original ancestral population into a number of different species with many different ways of life. Some species became large, the size of a large cat. Others evolved specialized teeth for eating, losing some teeth or adding cusps to the back teeth for grinding. Others evolved specializations of their limbs for moving about, mostly involving the ability to grasp tree branches with the hands and feet. Others became gliders or flyers. Some scientists believe that the living dermopterans, or "flying lemurs," as well as the megabats, or "flying foxes," are remnants of this "flying primate" radiation. The bulk of the archaic primate radiation was, however, confined to earthbound small-mammal niches in the trees.

Our ancestors among these earliest primates were the plesiadapiforms—generalized, small-bodied, and arboreal. As successful as they were in their time, their numbers began to wane in the Eocene, and by the end of this epoch, some 35 million years ago, they were extinct. The oxygen isotope record shows that global temperatures began a long slide downward beginning about 58 million years ago. The tropical forests of the archaic primates, particularly in those parts of their range farthest from the equator (North America and Eurasia), were affected. We know few of the details, but whatever changes in food resources and competitive interactions with other species occurred, they resulted in a major change of trajectory of primate evolution. By the beginning of the Eocene epoch some of the plesiadapiforms had evolved into prosimians—"primates of modern aspect" and the most primitive members of the order still alive today.

THE EMERGENCE OF PROSIMIANS

The earliest primates of modern aspect that we know of are set apart from their more primitive ancestors by several important traits. Some of these we know about because of fossil discoveries, and some we deduce from the comparative anatomy of modern primates. The eyes move close together on the front of the face, giving prosimians overlapping fields of vision that allow very accurate depth perception. The ends of the digits of the hands become flattened into sensitive fingertip pads supported by fingernails instead of claws. The brain enlarges in overall size. Particularly, the visual cortex, that part of the cerebrum located at the back (occipital) pole of the brain, enlarges, and the ancient smell-brain (olfactory cortex) decreases in relative size and importance. The jaw becomes heavier and connected by fused bone in the chin region in order to anchor teeth that are exerting more force in chewing harder foods. And fossil limb bones indicate that some species made a transition from the typical mammalian four-footed, head-down stance in climbing to one in which the hind limbs supported most of the weight of the trunk and body, which were held vertically. The prosimians thus evolved the ability to hold onto large vertical tree trunks in feeding, and some became extremely adept jumpers.

Why these changes occurred and in what exact sequence they appeared have been debated for nigh on a century. Early researchers thought that somehow there was an inexorable "trend" toward larger brain size in primates, culminating in the greatly enlarged cerebrum of humans. They tended therefore to posit brain size as the "prime mover" of primate evolution. But there is nothing intrinsic to brain size increase that would have driven primate evolution. Rather, the brain responds to natural selection as do all other structures in the body, and it is to those selective forces that we must look if we are to understand the emergence of the modern primates.

AN ORDER OF OMNIVORES

Anthropologists Robert Harding and Geza Teleki, in an exhaustive compilation of primate dietary patterns, concluded that primates were, across the board, omnivorous. Only in isolated cases did some species, such as colobus monkeys, become obligate herbivores, or other species, such as the needle-clawed galago, become dedicated gum-eaters. The primates as a whole have retained their generalized dietary preferences.

But primates are omnivores with a difference. They probably had origins as predator species, unlike such other familiar omnivores as pigs or rodents. This idea comes from Matt Cartmill of Duke University. Cartmill points out that somewhere along the line leading from the tree shrew–like ancestors of primates to the first true prosimian primates, the eyes moved from the sides of the head to the front. The old idea was that this forward position of the eyes served an important adaptive function in depth perception of tree-living animals. The visual fields of the two eyes overlap and thus slightly different angles of view are transmitted to the brain. This stereoscopic adaptation allows an accurate estimate of distance and depth of an object, such as a tree limb, to be judged before the animal leaps into thin air.

Cartmill believes that there was more to this story. He emphasizes the similarity between primates' eyes and those of such nocturnal predators as cats and owls, while he points out that many small mammals live in trees and do not *per force* have to evolve stereoscopic vision to do so. To catch small prey at night or at least in reduced light conditions in a potentially dangerous three-dimensional environment requires close-in focusing on prey and great accuracy in attack. Cartmill thought that the first true primates retained their insectivore ancestors' taste for bugs and small animals (and their spiky teeth betray this dietary preference), but that they became much more adept hunters than their walleyed predecessors.

The primates, however, have a unique way of dispatching their prey, and this aspect of basic primate adaptation does not fit Cartmill's scenario so well. Since time immemorial the mouth has been the organ of feeding, and most predator species have evolved specializations of the mouth region, such as a long, sticky tongue as in chameleons, frogs, or anteaters, or sharp, piercing teeth and strong jaws as in lions, crocodiles, or sharks. The primates lack such specializations and their hunting technique is remarkably primitive—they simply grab their prey, stick it in their mouth, chew it up a little bit, and swallow it. The primates even lack claws, which many other carnivorous species have so they can utilize their forelimbs in feeding. What is most telling for Cartmills's hypothesis is that claws are primitive for the mammals. If primates had first evolved as obligatory predators, claws would have been retained as very useful in helping to catch, pierce, and hold onto small prey. It follows that the flattened fingernailed digits of primates had to have evolved for specific adaptive uses probably unrelated to predation.

Robert Sussman and Peter Raven, in a unique collaboration between an anthropologist and a botanist, put forward an ecological argument for the evolutionary origins of the early primates. Noting that these species were forest-living and arboreal, they made the now obvious connection between the appearance of diverse flowering trees (angiosperms) and the opening up of new niches for evolving primates. In their scenario, primates became important dispersal agents for the new plants, eating their fruits and spreading the seeds via their droppings. The new plants in turn became important new food resources for primates. This hypothesis helps to explain the evolution of a tactilely sensitive organ like the primate hand, with its fleshy and ridged fingertips, designed for careful selection and deliberate removal of ripe fruit from branches rather than rapid dispatch of a moving prey target. A feeding adaptation including fruit also helps to explain why primates all have color vision, not an important adaptation in a nocturnal predator.

Cartmill's and Sussmann and Raven's hypotheses both explain important aspects of primate adaptations, and although they have been considered competing hypotheses, it is also possible that they are both partially right. Since primates are not now and probably never were either exclusively predatory or frugivorous, their

anatomical specializations may well reflect selective forces for om-
nivory. Indeed some primate field researchers have pointed out
that climbing onto the ends of branches for the best fruit and,
while there, picking off a few tasty bits of protein on the wing constitutes a very effective adaptation in a tropical forest environment. The fossil record is still too fragmentary to reconstruct
whether predaceousness or fruit-eating came first in prosimian
evolution, but they both must be ancient components in an overall
omnivorous dietary adaptation.

Prosimian Paradise

There is only one place on earth where a fauna has survived with
primates of analogous evolutionary grade to the prosimian world
of Paleocene North America. That is the large, equatorial island
of Madagascar off the east coast of Africa. Here a variety of
species of lemurs—strangely intelligent catlike and rodentlike animals—occupy habitats all over the island, mostly in the forests.
Some are nocturnal, but many are active in the daytime as well,
for here the rodents failed to colonize. Continental drift pushed
Madagascar out to sea, far enough away from the African mainland to keep the primates' voracious and rapidly reproducing
bucktoothed relatives off the island. Even more importantly from
a standpoint of lemur ecological competition, because many
prosimian lineages had increased in size to be much larger than
most rodents, no advanced primates—monkeys or apes—made it
to Madagascar either.

Several important aspects about the setting of Madagascar,
other than its isolation by continental drift, can tell us about the
ecological adaptations of the early prosimians. First of all, it is and
has been for all the time that prosimians have occupied it, near
the equator. Although some latter-day prosimian species have
spread to some of the drier parts of the Madagascan forests, the
seasonal and temperature variations on Madagascar are quite mild

compared to higher latititudes. Rainfall varies with the monsoons blowing in from the Indian Ocean but temperatures are relatively constant. Not only do oceanic currents buffer the island from major temperature change but insolation (the amount of incident sunlight) and day length stay relatively constant throughout the year. This is because the effect of the earth's tilt is least felt near the equator. Constancy of environment is the key to understanding why a prosimian world has been able to persist on Madagascar and nowhere else in the world. And Madagascar is large enough to have preserved some of the diversity of the prosimian evolutionary radiation, from tiny dwarf lemurs to the dog-sized and terrestrial ring-tailed lemurs. Some species, such as *Megaladapis,* became huge before their gorilla-sized bulk and unsuspecting temperaments made them easy targets for extinction by early human colonizers of the island.

The constancy of their environment and their lack of ecological competitors has allowed a degree of dietary specialization among the Madagascan prosimians that is reflective of the kinds of dietary specializations of the ancient prosimians. The lemurs as a group are picky eaters. For example, in the largest captive colony of lemurs in the world, at the Duke University Primate Center in North Carolina, lemur food stores of fresh mango leaves have to be flown in every day from Florida. It is very important that the leaves are still *on the branches.* Otherwise, the largest of the lemurs, the indris, will turn up their noses at them. The founder of the Duke facility, John Buettner-Janusch, discovered that all the previous attempts to breed the endangered indris in captivity had failed because they simply could not pick and then eat their favorite food leaves. When he took single mango leaves and tied them onto an artificial tree so that the indris could actually pluck them off, as they do in the wild, they ate! If most North American Paleogene (Paleocene and Eocene) prosimians were such prima donnas when it came to food, it is not hard to see how any change in their environment and food source would have caused a problem.

THE ASCENDANCY OF THE RODENTS AND DEMISE OF THE HIGH-LATITUDE PRIMATES

All of the North American small mammals with primatelike adaptations are rodents of one sort or another—gophers, squirrels, marmots, pack rats, chipmunks, and field mice. Rodents are a remarkably successful group of mammals. In North America they replaced the primates, who died out some 45 million years ago, not to be represented again until huge, bipedal primates called humans made the trek from Asia into the New World, only a few thousand years ago. What do the rodents have that the primates lack? The primates are a lot smarter, although people who have only barely been able to wrest control of their house away from a family of mice or rats might question that assumption. Usually another species, a domesticated small carnivore (a cat), has to be brought in by the humans to finish the job. Rodent–primate competition over resources is an ancient one, and humans usually win only because they are much bigger and are able with their superior intellectual capabilities to devise new and innovative ways to do in the rodents. This is why "building a better mousetrap" will cause the world to beat a path to your door. But despite the accolades of our fellow humans and the success of some of our inventions, the rodents continue to thrive. We primates have to be content with keeping them at bay, out of our immediate sight, and out of our food supply.

Earlier primates did not have at their disposal either the advantages of great size or of vast intelligence in their ecological competition with rodents. Many lineages lost that competition, which explains why the view from my window does not reveal a riot of primate diversity. Compared to rodents, primates put a lot more parental effort into raising their offspring, helping them to survive, and because of this they have fewer offspring per litter. Rodents, in contrast, generally produce many more offspring, and they are willing, evolutionarily speaking, to sacrifice a large number of them to the forces of natural selection. Untold millions of

owls, hawks, and small carnivores have gorged themselves on billions of young rodents over the past 50 million years in the Oregon outback, but still their population numbers are high and they
are successful species. The reproductive potential of rodents is
phenomenal, allowing rodents to not only sustain tremendous
predation but also to outcompete other species in the ecosystem
by outbreeding them.

Evolutionary biologists have referred to the modes of natural
selection that rodent and primate evolution exemplify as *r selection* and *K selection,* respectively. The terms come from population genetic equations describing the interactions of birth rates,
predation rates, and other ecological parameters. Generally speaking, r-selected species are those with great reproductive potential,
capable of producing several large litters over a period of a year's
time, usually in seasonal peaks correlated with the abundant food
supply. They suffer high predation pressure and high death rates.
But r-selected species such as rodents shed no tears for their lost
offspring. They just produce more.

K-selected species are very different. They have small litters.
All primate females, for example, have only two nipples for suck-
ling one or two offspring at a time, indicating that large litters
have not been a part of the primate adaptation for tens of millions
of years. K-selected species keep their unborn offspring developing longer inside the mother's body before birth, accounting for
lower reproductive rates but a higher chance for individual survival after birth. The primates even enhance the quality of life for
their unborn offspring. The primate placenta—the flat, blood-
filled connection plate of the fetus's umbilical cord to the mother's
uterus—has a thinner lining than other mammals, thus allowing
better delivery of maternal oxygen and nutrients to the fetus and
better removal of fetal waste products. Primate parents, usually
mothers, continue this coddling of their infants after birth, ensuring that they have sufficient food, warmth, and protection for
their survival. When offspring of a K-selected species die, all of
that invested energy is lost, entailing a high cost to the species. But

if the parental investment is effective, many fewer K-selected off-spring die and their species thrive. If this were always the case, the early primates would not have died out in North America. What happened?

A COLLISION OF REPRODUCTIVE STRATEGIES AND ENVIRONMENTAL CHANGE

Why species went extinct in the past has always been more than a little mysterious. We know so little critical detail about the past—the organisms and their adaptations, the terrain and its habitats, the climate and its perturbations. And our experience with extinction has been strongly colored by the rapid elimination of numerous species by the combined human effect of "civilization." It is easy for us to imagine a catastrophe, analogous to human slaughter of Steller's sea cow or widespread despoliation of the riverine environments of the snail darter, as responsible for wiping out species in the past. Some scientists subscribe to this "catastrophic" view of evolutionary history and posit such events as meteoritic collisions with earth, viral epidemics, and explosive evolutionary changes as responsible for species extinctions in the past.

Most of these "unitary cause" arguments for extinction are unconvincing. First of all, natural events drastic enough to wipe out species, that is, killing *all* breeding members of a population, must be extremely rare. Consequently, they require a feat of imaginative reconstruction that, without strong empirical evidence, is difficult to make. True, certain cataclysms, such as the drying up of the Mediterranean Sea, did occur. But even in this case, as we shall see, accompanying massive species extinctions do not seem to have occurred at this same time. Second, whatever catastrophe is invoked to explain extinction must have been selective—it must have killed off one species or one group of species but not others. The meteorite that theoretically hit earth at the end of the

Mesozoic era, for example, had to have killed off giant carnosaurs, tiny dinosaurs, and sea-living plesiosaurs, but not have harmed any of the other species living in these same environments.

Extinctions in the past, one may reasonably surmise, are usually results of combinations of species and environmental interactions. The environment changes, shifting food resources around, and these changes create stresses on some species while opening up opportunities for others. Species may compete in many ways— for space, for nesting or sleeping sites, for shade, for warmth, for water, for anything that they need that may be in short supply— but the most common arena of competition is likely what they eat. Dietary competition between species happens day in and day out, and observations on living species show that it can be exacerbated during certain annual seasons of scarcity. Animals with similar dietary adaptations are most likely to have competed, and both primates and rodents, as a general rule, are omnivores. It is thus reasonable to expect that they would have been ecological competitors.

The Cenozoic era began with a series of global cooling events that had widespread environmental effects. There are many who consider that these changes were the very ones that slowly did in the dinosaurs. The warm-blooded birds and mammals took over an increasingly changeable world. Forests became cooler and less swampy, and the numbers and kinds of fruiting trees increased. The primates probably developed both as predaceous small animal-eaters and fruit-eaters. But as climates became progressively cooler and drier, fruit trees became sparser. They, along with other plants, evolved in response to the changing climate. Lush tropical fruits gave way to thick-coated fruits and nuts, more impervious to drying out. Hard-covered seeds, nuts, and grains became more the norm, and primate adaptations were less up to the challenge. Still today we can easily eat soft fruits such as bananas, oranges, apples, dates, and peaches with only our hands, but we require tools, such as a nutcracker, to eat hazelnuts, brazil nuts, and walnuts, and a millstone or fire to eat grains such as wheat, barley, and rice. The rodents with their formidable, ever-growing front incisors could eat through the toughest nut and grain seed coats and successfully ingest foods that non-tool-using primates could only dream about. And they could outbreed primates by a wide margin. It is no wonder that at high latitudes where climatic changes would have been greatest, primates went extinct by the end of the Eocene epoch. The rodents were very likely one of the prime biotic agents responsible.

In North America, Europe, and northern Eurasia, the rodents took over many of the primate niches. They themselves underwent an adaptive radiation, their species spreading and differentiating into many patterns of adaptation, but they did not *become* primates. Squirrels, for example, became denizens of the trees, like the early primates. But their eyes stayed on the sides of their heads, their sense of smell remained an important part of their adaptation, they retained their claws, and their brains stayed relatively small. Because squirrels do not have accurate stereoscopic vision, they have a parachute-like tail that breaks their relatively frequent falls out of the trees. Because they do not have grasping hands with an independently moving, "opposable" thumb, they must use both hands in eating their food. And they do not supplement their nut, seed, and fruit diet with any animal protein.

Nonhuman primates were gone from North America by the beginning of the Oligocene epoch, some 34 million years ago, never to return except by the hand of man. We can surmise that they were done in by a combination of the marked climatic shift toward cooler and drier conditions that occurred at this time (see Figure 2), associated changes in vegetation that altered their multistoried forest habitats, and ecological competition from the advancing and differentiating rodents. Perhaps the distances between trees just became too far to leap. Or the fruits and insects that they fed on became too rare to sustain them. The rodents perhaps beat them to the best fruiting trees and were able to eat the fruits before they were quite ripe, gnawing through the still tough

Figure 2

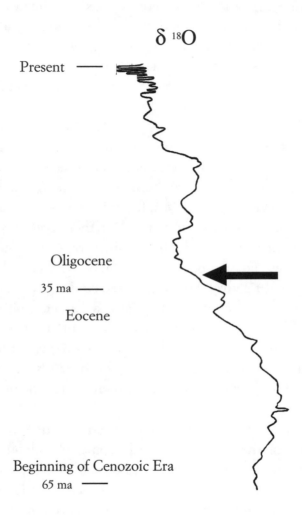

$\delta\ ^{18}O$

Present ——

Oligocene

35 ma ——

Eocene

Beginning of Cenozoic Era

65 ma ——

The arrow indicates the major downturn in global temperatures that accompanied the beginning of the Oligocene epoch. The important primate fossil site of the Fayum, Egypt, records this time period.

rinds before the primates could. Whatever the exact combination of factors that wrought the extinction of North American prosimians, the same underlying paleoenvironmental changes effected the emergence of true higher primates, the anthropoids, in other parts of the world.

THE OLIGOCENE AND THE FIRST LARGE PRIMATES

In the Tropics, the climatic changes that occurred at the beginning of the Oligocene epoch, 34 million years ago, were not so markedly felt as they were at higher latitudes. Forest environments changed in the composition of their species in response to climatic cooling, lower humidity, and lower precipitation, but they nevertheless retained their multistoried complexity. The forest-tied primates with their K-selected adaptations continued to thrive.

Much of our knowledge of this time in primate evolution comes from Africa—from one remarkable site in the Sahara Desert of northern Egypt. Despite this solitary ray of light, we are able to make some deductions about general primate distributions from other data. Because continental drift cut off the archaic primates of North America and Eurasia from the ancient southern continent of Gondwanaland, prosimians never reached South America, Australia, and Antarctica, the modern continents formed from this old supercontinent. Here primitive pouched marsupial mammals continued to hold sway over the small-animal niches in the trees. The original Antarctic fauna was frozen into extinction as that continent separated from Australia. Australia stayed isolated by its deep and wide South Pacific moat and thus retained its diverse marsupial mammal diversity. South America also was a marsupial stronghold, lacking primates, until it drifted northward and modern placental mammals flooded across the Isthmus of Panama. Prior to contact, there was even an Argentinean platypus, an egg-laying mammal even more primitive than marsupials. Primates were to make it to the lush South American rain forests, but they were late immigrants.

Thus, by the Oligocene, the scene of primate evolution shifts to what we rather incongruously call the "Old World"—that part of the Northern Hemisphere consisting of the contiguous landmass of Eurasia and a major piece of Gondwanaland—Africa—that drifted northward to meet it. Of course there is nothing intrinsically "old" about it. It just happens to be what our intellectual forebears had imperfectly known about and put on maps before 1492.

Evidence now indicates that the earliest "higher primates," anthropoids, appeared as early as the middle Eocene epoch. Sites in China, Burma, and Morocco now show that primates with such humanlike characteristics as forward-facing eye sockets totally enclosed by bone, well-developed jaws with fused bone in the chin region, and single frontal bones in the forehead had evolved by 40 million years ago. Since no primate like this is known from North America, we can presume that faunal interchange had been cut by this point between Eurasia and North America. Contact with Africa, which we may probably assume had a marsupial-dominated fauna, allowed Eurasian prosimians to extend their range into the tropical forests of Africa for the first time. Because of the tropical conditions, there was continuous forest from Africa to Eurasia, and thus Morocco, Burma, and China would have been much more ecologically similar than they are today. The primate faunas thus could be expected to have been similar throughout this region of the Old World.

The site of the Fayum, in Egypt, is situated within this ancient Old World forest zone, and it is replete with primates. Elwyn Simons, the world's senior paleoprimatologist, has worked at the Fayum for two decades, and has succeeded in uncovering a primate radiation of impressive diversity. It has been for several decades the primary window onto this period of florescence of the higher primates, but recent discoveries by Simons and his team from Duke University continue to provide new insights.

At the time that the Fayum deposits were laid down by ancient channels of the Nile, Egypt lay several degrees south of where it is now. This more southerly location, as well as the proximity of moisture-bearing winds from the large Tethys Sea, the much larger ancestor of the Mediterranean Sea to the north, made the Fayum an area of dense tropical forest. The lowest strata of the Fayum are Eocene in age, about 36 million years old, but most of the deposits are Oligocene in age, some 31 million years old. The Fayum preserves both prosimians and the first well-preserved anthropoids.

The striking aspect of the primate fauna that Simons has discovered is that all of the species are small. Among the prosimians, a recent discovery is the small lemurlike species *Plesiopithecus,* which had incisor teeth elongated to form a "tooth comb" for grooming, and which may provide a mainland African link with the Madagascan lemurs. Small, tarsier-like prosimians show that the Fayum primates shared close ties with the Asian rain forests, where tarsiers survive to this day.

Most interesting from the standpoint of human evolution is the appearance of two groups of diminutive higher primates, anthropoids, in the fossil record of the Fayum. One of these groups has the three premolars characteristic of the New World monkeys, a trait that is today no longer found among the primates of the Old World. This group, known as the parapithecids, probably accounts for the origin of the New World monkeys. As unlikely as it sounds, some of these tiny primates apparently dispersed to South America along with African hedgehog rodents, probably on floating rafts of vegetation that broke off from westerly flowing rivers, perhaps the Niger or the Zaire. Hurricane winds, which today originate off the West African coast and buffet the Caribbean and northern South America, could transport such a raft across a then-narrower Atlantic in a couple of weeks. When these small primates reached the tropical forests of South America they underwent a rapid radiation, the details of which primate paleontologists are just beginning to unravel.

The second anthropoid group at the Fayum is the propliopithecids, a group that includes our probable ancestor at such a distant time. The most well-known of this group is *Aegyptopithecus,* but Simons's recent discoveries have shown that, like the other primates at the Fayum, the earliest members of this family were tiny. A new species from the Eocene of the Fayum that Simons named *Arsinoea* in 1994 was no larger than a small chick!

The paleontologist Alfred Romer made the observation that in many evolving lineages of vertebrates there is a clear tendency to increase in body size over time. This generalization has become known as Romer's Law, and it accurately describes what happened to the propliopithecid lineage from the Eocene to the Oligocene. *Aegyptopithecus* evolved to become about the size of a large house cat, with males about twice as big as females. This was indeed a significant increase in size, and it represents the first time that anthropoids made the transition from "small animals" to "medium-sized" animals. Why this transition occurred in these particular primates involves a consideration of the general applicability of Romer's Law.

CLIMATE CHANGE AND THE ORIGINS OF THE ANTHROPOIDS

As conditions get colder, animals get bigger. A German naturalist noticed this back in 1837 and the generalization still bears his name—Bergmann's Principle. Bergmann compared body sizes of related animals, bears for example, from the equator to the poles. He found that as one progressed from sun bears to black bears to grizzly bears to polar bears, size increased. One explanation for Romer's Law then is that it is a special application of Bergmann's Principle through time—as conditions become colder over time, species evolve to be bigger. There is a simple principle of physics that explains body size increase with decreasing temperature. As mass of a body increases, its relative surface area decreases. Thus, with warm-blooded mammals, the bigger the animal the smaller the relative body surface that would be exposed to the cold.

The oxygen isotope curve shows that there was a significant period of global cooling at the beginning of the Oligocene epoch. It is tempting to jump to the conclusion that this decrease in temperature prompted the evolution of larger body size among the anthropoids. While it is not yet possible to rule this out as an explanation, there are other possibilities.

Interspecies competition can also lead to an increase in body size. This can happen as a result of "scramble competition," when individuals of one species directly compete for some environmental resource in limited supply—for example, meat from a kill or fruit from a tree. The larger animals usually displace the smaller, thus winning a competitive advantage in access to food resources. In the case of the early anthropoids, the rodents are again a prime candidate for the competitors. One species of Fayum primate, *Serapia eocaena,* for example, is the spitting image of a squirrel, if you do not look too closely at its teeth or hands.

Larger body size can evolve as a defense against predation. Rather than develop fleetness of foot, bony body armor, an offensive smell, or an especially effective way of hiding, animals can just grow too large for predators to handle. Imagine for a moment that you were one-sixth the size you are, say less than one foot tall. The number of potential predators in your everyday environment would be tremendously increased, beginning with your pet cat and dog. The early anthropoid's arboreal environment would have placed a limit on how large they could have become (as well as limiting the size of their potential arboreal predators). Up in the branches, the bigger you are the harder you fall. So the list of predators for the Fayum primates would have included animals whose means of locomotion included access to the trees—snakes, particularly booid constrictors, raptorial birds such as eagles, hawks, and owls, and any small carnivorous mammals that could climb.

Finally, evolutionary forces within the species can make for larger body size. Charles Darwin first noted that some changes in species could not be easily explained by adaptation to the environment. An adaptation like the elaborate ornamentation of a peacock's tail could only have served to attract predators, which would have eaten its possessor. This type of adaptation could only be explained from a standpoint of competition for mates among members of the same species. Darwin called this type of selection *sexual selection,* to distinguish it from the more common *natural selection.* Traits of a species evolved because either males were able to beat out other males for mating access to females, or because females chose certain males to mate with on the basis of those traits.

The increased sample of fossils from the Fayum now make it possible to sort out males from females in the species *Aegyptopithecus zeuxis,* the prime candidate among the Oligocene anthropoids to be our direct ancestor. It is apparent from a size comparison of the canine teeth, skulls, and limb bones that there was a marked size difference between males and females. Males were about one-quarter larger than females. This *sexual dimorphism,* as it is called, is a characteristic of many higher primates, and it speaks volumes on the social behavior and mating adaptations of these early anthropoids. Males clearly competed for females, and larger size (again limited by life in the trees) was a distinct advantage. Thus, sexual selection would have been a factor acting to increase body size, at least in the male anthropoids.

From what we know of the Oligocene primates, all of these factors—cooler temperatures, scramble competition, predator avoidance, and sexual selection—may have played a role in the increase in body size. We just do not know enough to posit which one was the most influential. We can be fairly certain, however, that the initiating cause, the "prime mover" that set the other forces into operation, was climate change. Not all evolutionary biologists would necessarily agree, and it is worthwhile to consider one other explanation of evolutionary change.

WHO RULES? THE RED QUEEN OR AEOLUS, GOD OF THE WINDS

In one of Alice's excursions through the looking glass, she encounters the Red Queen, who must keep running as fast as she can just to stay in the same place. Evolutionary biologist Leigh Van Valen at the University of Chicago used her as a metaphor to help explain evolutionary patterns. Species must be constantly changing to stay in the same adaptational or ecological place. Because predators are constantly being winnowed by natural selection to be better able to catch their prey, the prey species will be constantly selected to be better able to escape the predators. If a species is to just maintain its ecological position it will be constantly evolving. Van Valen termed this phenomenon the "Red Queen Hypothesis." It explains why evolution seems to progress at a somewhat constant rate through time.

If the Red Queen Hypothesis is correct, then is it necessary to postulate a formative role for environmental change in the evolutionary history of life? Evolution will happen anyway. Is it necessary to search for Aeolian capriciousness to create environmental catastrophe in order to account for evolutionary change?

As with most dichotomies, this one is largely unnecessary. Van Valen's model works well under conditions of either constant conditions or conditions that are changing at a relatively constant rate. Coevolution, fostered by complementary changes in evolving plant and animal communities, will take place in relatively stable environments, but it will also occur in suddenly changed environments. When the environment is the constant, we tend to ascribe evolutionary change to the Red Queen, but when sudden environmental change outruns the Red Queen or requires her to run faster than she normally would, we ascribe evolutionary change to the environment. Thus, the effects of the Red Queen and Aeolus intergrade to account for the slow and gradual changes in evolution on the one hand, and precipitous changes on the other.

This way of looking at evolutionary change is probably more

reasonable than regarding the history of life as long periods of stasis automatically punctuated by periods of rapid change, a model called "punctuated equilibrium" by Stephen Jay Gould and Niles Eldredge. Punctuated equilibrium was an attempted description of paleontological history, mainly of invertebrates, based on Gould and Eldredge's reading of the fossil record and their comparison of their understanding to sociopolitical history. Most evolutionary biologists did not care for the rather mysterious vitalistic wonder of it all, and found it too much to ponder what underlying cosmic laws might account for both the fall of the ancient Egyptian empire and the demise of the ammonites. Some good paleobiology was, however, accomplished in the effort to test the model, and for this we can be grateful. But punctuated equilibrium as a predictive model is useless since it lacks any clear causative association in nature.

A Riot of Apes in the Miocene

With the increase in body size and dominion established over the tropical forests, the primates emerged from the Oligocene, some 20 million years ago, with an abundant and wide range of species (see Figure 3). We know most about these species from sites in western and central Kenya, and from sites in eastern Uganda. They are called apes—proconsulid apes.

The early apes were in general smaller and much more monkeylike than what we now think of as apes. Their skeletons show that they were quadrupeds, capable of climbing, rather than the specialized arm swingers and branch hangers that modern apes are. They came in all sizes and shapes, from the gorilla-sized *Proconsul major* and *Turkanapithecus heseloni* to the diminutive *Micropithecus clarki*. Some of the smaller varieties, such as *Dendropithecus macinnesi,* were beginning to specialize in arm swinging and were akin to the adaptation seen today in the South American spider monkey. Somewhere in the small-to-middle-size range of the proconsulid apes lies our ancestor—actually the common ancestor of the living apes and humans. It may be a species close to the relatively well-known species *Proconsul africanus,* a tailless climber the size of a small dog.

The forests in which the proconsulids lived was dense but far different from the rain forests of the Oligocene and preceding epochs. The early Miocene proconsulid site of Koru in western Kenya preserves the remains of a mature forest but one that was adapted to periodic dry conditions. The site of Napak in eastern Uganda is perhaps the driest of these early Miocene sites, and although a forest, it probably had open patches within it. Most of the fossils here were covered by an ash that erupted from a nearby volcano.

THE APES LEAVE AFRICA

As Africa drifted northward at the beginning of the Miocene, a broader land connection was established between the Arabian Peninsula, primitively part of Africa, and the mainland part of Southwest Asia. From some newly discovered faunas in Saudi Arabia we know that this part of the world was densely forested in the Miocene, and that apes inhabited what now is the totally treeless Empty Quarter. The Sahara was still largely a forest, although the first plant fossils indicative of aridity come from this time in southern Algeria. Forest stretched from Africa eastward to southeastern Asia and westward to Europe. With the newly established land connection, elephants, originating from Africa, spread into Eurasia for the first time. Apes came with them.

Exactly how many and what kinds of apes made it out of Africa are still questions to be answered. The ancestor of the gibbon, the living Southeast Asia long-armed tree-swinger, may have already split from the line leading to the larger apes before the African ape diaspora. Or the split may have occurred just after the spread of apes to Eurasia. In any event, by the middle Miocene, an ape in

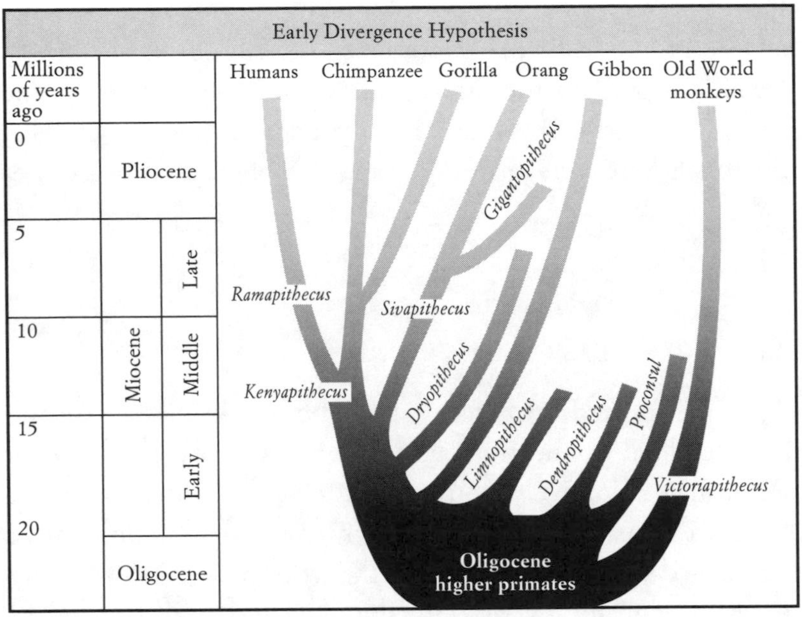

Early Divergence Hypothesis

Millions of years ago			Humans	Chimpanzee	Gorilla	Orang	Gibbon	Old World monkeys
0	Pliocene							
5	Miocene	Late	*Ramapithecus*					
10	Miocene	Middle	*Kenyapithecus*	*Sivapithecus*	*Gigantopithecus*			
15	Miocene	Early		*Dryopithecus*	*Limnopithecus*	*Dendropithecus*	*Proconsul*	*Victoriapithecus*
20	Oligocene							

Oligocene higher primates

Figure 3: *Theories of hominid evolution. The older paleontological view that placed humans far outside the context of primate evolution, exemplified by the "Early Divergence Hypothesis," has now*

Late Divergence Hypothesis

| Millions of years ago | | | Humans | Chimpanzee | Gorilla | Orang | Gibbon | Old World monkeys |

Gigantopithecus

Sivapithecus/ Ramapithecus

Kenyapithecus

Dryopithecus

Afropithecus

Limnopithecus

Micropithecus

Dendropithecus

Proconsul

Victoriapithecus

Oligocene higher primates

been replaced with a view that emphasizes how close human ancestry is to that of our living African ape relatives, the "Late Divergence Hypothesis."

China, *Dionysiopithecus,* records what most primate paleontologists consider a good fossil ancestor for the living gibbons.

A very strange ape named *Oreopithecus*—one with the body of an orangutan but the teeth of an overgrown prosimian—dispersed into Europe from an early Miocene ancestor known from Kenya. It lived on into the late Miocene in the coal forests of central Italy, where it eventually died out as its habitats disappeared. Most of the apes of Europe were dryopithecids, small to medium-sized species that resembled their proconsulid ancestors. They were also forest dwellers, and like *Oreopithecus,* they went extinct in the late Miocene by the time the European forests thinned out too much to support them.

THE RISE OF THE MONKEYS

In addition to habitat change, there is another explanation for the passing of the great diversity of apes seen in the early and middle Miocene. It is the evolutionary diversification of a related group of primates, the monkeys. Monkeys are in general smaller than living apes, they usually have tails, their molar teeth are pushed up into transverse crests called lophs, and their brains are relatively smaller than apes' brains. The distinction between monkeys and apes is still muddled in the popular mind: Chimpanzees, gorillas, and orangutans are still frequently called "monkeys," and very good monkeys, such as Barbary "apes" and Celebes black "apes," continue to carry their misnomers.

The evolutionary distinction between apes and monkeys is razor-sharp, however. The earliest monkeys are singular species with lophed molar teeth that appear in the early Miocene. They must have lived surrounded by trees teeming with different species of early apes. The apes had dominated the primate ecological spectrum since the Oligocene, when they and the monkeys had last shared a common ancestor. The apes had won the first round, but by the middle to late Miocene it was the monkeys' turn.

Although we would like to have more sites to more fully document the transition, the available evidence shows a clear inverse relationship between number of species of apes and monkeys through time. As monkeys increase in diversity, apes decline. The similarities in their adaptations, as indicated by skeletal form, make it likely that the decline of the apes can be directly attributed to the rise of the more r-selected, less intelligent monkeys. The monkeys could perhaps subsist on diets of less high-quality fruit and protein, which became sparser and sparser as the Miocene progressed. They successfully radiated into a number of niches, taking over ape habitats and spreading throughout the Old World. The even more impressive radiation of monkeys in South America happened in the absence of apes, who were perhaps already too big in the Oligocene to have made the trans-Atlantic journey with them and their rodent companions to colonize the New World.

THE EVOLUTIONARY ORIGIN OF THE ORANGUTAN

The middle Miocene in both Africa and Eurasia first records the presence of grassland and open wooded savanna in the middle Miocene, 13 to 15 million years ago. This is the time that protein and DNA (deoxyribonucleic acid) analyses of living apes and humans indicate that the orangutan, the only extant Asian great ape, split off from its African cousins, the chimpanzee, gorilla, and humans. The timing corresponds well with the cutting of forest connections between Africa and Eurasia.

The European apes, the dryopithecids and *Oreopithecus,* continued to live in their forested cul-de-sac. They were cut off from further contact with ape populations to the east and to the south. The spread of open woodlands in southernmost Europe and southwestern Asia through the middle to late Miocene heralded the appearance of faunas in which primates played almost no part. Sites such as Maragheh in Iran and Samos and Pikermi in Greece preserve beautifully diverse animal communities, but there are no primates, except for some very fragmentary remains of monkeys. The Mediterranean crossroads of faunal migration became essentially closed to apes.

In the rainier and more well-wooded parts of Asia, to the east, some apes persisted. The woodlands surrounding the mountain massifs of the Zagros in Turkey and the Himalayas allowed a group of apes known as sivapithecids to survive. One of this group persisted as the orangutan, up until only a few thousand years ago still inhabiting the Chinese and Vietnamese mainland, but now limited to only relictual island populations in Indonesia (Borneo and Sumatra). Pressed into this same Southeast Asian forest isolate are the only other extra-African group of surviving apes, the gibbons and related siamangs.

Our realization, born of molecular and fossil discoveries over the past twenty years, that the orangutan bears no intimate relationship to human or African great ape ancestry allows us to concentrate the search for earliest hominid origins on the African continent. Eurasia in fact has a much better fossil record of the late Miocene than does Africa, a fact fostered no doubt by a century-and-a-half-old belief among many paleontologists that human origins lay in the Northern Hemisphere. They did not. Africa, the part of Gondwanaland that broke off early from the other southern continents and first docked with Eurasia, provided a haven for primates that had originally arisen in the north. Climatic change and interspecific competition drove North American primates to extinction and pushed Eurasian primates southward. It was Africa that shaped anthropoid evolution for the next 40 million years, and it is to a consideration of this continent's latter ecological history and ape evolution that we will turn next.

2

The Earth Cools and the Gorilla Evolves in Montane Isolation

There are a few national parks in the high lake country of northwest Rwanda, southwest Uganda, and eastern Zaire where one can trek into the montane forest habitat of the mountain gorilla, its last natural refuge. There it is possible to visit, after a strenuous hike of several hours, gorilla family groups living in the wild and habituated to the presence of humans. There are no bars, no glass enclosures, no water barriers—just you, the gorillas, and a few dozen feet of space in between. It is an unforgettable experience.

One is acutely aware that this is not a human environment. It is cold—a kind of wet, cloud-shrouded chill that settles deep into your bones early in the morning and stays there. Your muscles ache from the constant shivering. It is also an environment in which it is impossible to walk. In most places you cannot see the actual surface of the ground—it is covered in a mass of tangled roots, branches, moss, and leaves. You slip, crash, stumble, and crawl over and through this substrate, stirring up an immense amount of noise in the quiet forest. Unhabituated gorillas easily hear humans approaching and glide effortlessly and gracefully away to a safe distance. They seem to move slowly and methodically, carefully placing their hands and feet, but then they are gone, impossible to catch. But the gorillas that have become habituated, through the patient work of dedicated primatologists and conservationists, largely ignore the humans who every day barge into their environment.

Sitting and quietly observing a group of gorillas feeding, which is largely what these huge primates do, one is struck by how powerful they are. The chewing motions of an adult gorilla entail movement of the entire head, as the heavy muscles that unsheathe the skull and move the jaws contract. As a gorilla nonchalantly reaches out to bend down a bamboo tree, with arms the girth of an ice skater's legs, the cracking sound reverberates through the forest. Using their canines in a sideways bite, gorillas lever open the trunks of these pithy trees to get at the soft interior. It is impossible for a relatively well-muscled, six-foot-two-inch human male to even budge a similar-sized tree, and human teeth simply bounce off the hard exterior. I tried.

It is easy to see why early explorers and hunters perceived the gorilla as a monster, as ferocious as it was terrifying. The gorilla's strength and large bulk alone make it a creature intimidating to humans. But as primatologists in the last thirty years have discovered, gorillas are far from aggressive animals. Unlike brash and loquacious humans and chimps, they have quiet, unassuming, and even morose personalities. Sex, which is such a preoccupation with chimps and humans, is not a big deal for gorillas. Copulations are almost never observed.

I remember sitting on my haunches half hidden by vegetation, watching the gorillas and marveling at how little I, or anyone for that matter, knew of this unique primate. Darwin, in one of the most oft-quoted one-liners in anthropological history, had correctly predicted in 1873 that human origins lay in Africa, somehow connected to a shared ancestry with the gorilla and chimpanzee. But how? And when? The reality of this powerfully impressive primate in its natural habitat was in front of me. Yet very little progress had been made in understanding its evolution.

In 1982 I started the Western Rift Research Expedition, which has worked in fossil deposits in eastern Zaire and western Uganda—gorilla country. This research, and that of our colleagues, has for the first time contributed to concrete knowledge of the gorilla's historical biogeography—where gorillas are found and why. Fossils complete enough to inform us of early gorillas' adaptations have so far eluded us, but establishing the contexts of gorilla evolution can be considered of almost equal importance at this stage of our ignorance. The thesis of this chapter is that the environment of the gorilla is the actual cause of its evolutionary separation from its closest primate relatives, human beings and chimpanzees.

THE (NON)FOSSIL RECORD OF GORILLA EVOLUTION

In the days of the Early Divergence Hypothesis, most paleoanthropologists thought that there was a good early Miocene ancestor for the gorilla. The gorilla-sized *Proconsul major,* represented by a partial facial skeleton and found in western Uganda, was considered to be in the right place and at the right time to be a believable gorilla ancestor. But for reasons discussed previously, particularly the increasingly precise molecular record of ape evolution and the growing fossil record of Eurasian apes, it soon became apparent that *Proconsul major* as well as some other large Miocene apes in Africa were not likely to be the documentation of the gorilla lineage that many had hoped for. Instead, these animals were earlier parallelisms, mirroring in some ways the large extant African ape that we have termed the gorilla, but in actuality not closely related to its ancestry.

One of the most convincing arguments against the possible

gorilla ancestry of large Miocene apes is that they did not share the climbing and hanging adaptations that typify modern African hominoids. *Proconsul major,* for example, would not have been able to habitually raise its arms straight above its head and suspend its weight from an overhanging branch. Rather, it would have climbed in branches or moved around on the ground more like a gigantic baboon, with its hands firmly planted, palm-down, on a subhorizontal tree trunk or on the ground. Gorillas, in contrast, have heavily developed flexor muscles of the forearms that do not allow their fingers to fully flatten out in extension, and they thus keep them curled up when they "knuckle-walk" on the ground.

If large-sized proconsulids dating to the early Miocene are not the ancestors of the gorilla, are there any possible fossil ancestors in the late or middle Miocene? Professor Ishida of the University of Tokyo discovered what he thinks is a possible gorilla ancestor in northern Kenya, at the site of Samburu Hills. The fossil from this site is a large mandible, but there is little in addition to its size that uniquely relates it to the gorilla. Meave Leakey of the Kenya National Museum discovered another large late-Miocene ape in Kenya that she named *Afropithecus,* but her anatomical study of the facial skeleton of this fossil also showed no unique connections with living gorillas.

Even in the Pliocene and Pleistocene there are no fossils that can be definitively placed on the gorilla's family tree. An expedition to western Uganda led by Martin Pickford and Brigitte Senut of the Museum of Natural History, Paris, turned up a single large but very eroded canine tooth from a site called Nkondo about 3.5 million years old. When Pickford and a colleague examined the specimen microscopically its enamel looked hominoid-like. They guessed that it was a gorilla. But even if it is, the fossil is so fragmentary that it does little to advance our understanding of the anatomical changes that constituted gorilla evolution. Nevertheless, it is the sum total of the fossil record uniquely leading to the gorilla.

Considering the sorry state of the gorilla's fossil record, is there anything we can say about where these hominoids came from? I believe the answer is yes. Reliable data come from the modern biogeographic distribution of gorillas, their molecular evolution, and the paleontology of sites that document the environmental history of the areas of gorilla evolutionary diversification. These are contextual answers. They come directly from the multidisciplinary approach of anthropogeny. Perhaps if this search into the gorilla's dim past were a murder investigation we might say that all the evidence is circumstantial. Only with further research we will find the "smoking gun"—the definitive fossil skulls, mandibles, and skeletal elements that will test the hypotheses outlined below.

GORILLA MOLECULES AND THEIR EVOLUTION

The first attempts at estimating the date of the "ape–human" split by using biomolecular data were made by Allan Wilson and Vincent Sarich of the University of California, Berkeley. Wilson and Sarich in 1967 reported results that indicated a three-way divergence of between 4 and 6 million years ago for gorilla, chimpanzee, and human. In the thirty years that have elapsed since this pioneering work, many biomolecular anthropogenists have investigated the problem. They have attempted to tease apart this evolutionary trichotomy and resolve which of the evolutionary divergences occurred first, and to refine the dating of the events. Since molecules evolve at different rates, any one of which may or may not correspond to the evolutionary rates of the animals' morphology and behavior, a number of molecules, including DNA, have now been studied.

Charles Sibley and Jon Ahlquist of Yale University in 1987 succeeded in showing to most researchers' satisfaction that the gorilla diverged first from the common human–chimp ancestral line. These researchers used a technique that paired up strands of DNA from different species, then measured how much heat it

took to break them apart again. The closer the genetic relationship between species, the stronger the chemical bonds between
the strands, and thus the higher the temperature needed to "melt"
the strands apart. Gorilla–human and gorilla–chimp DNA hy-
brids melted at a lower temperature than did chimp–human hy-
brids. Chimps and humans are therefore closer genetically to each
other than either is to the gorilla. Most molecular biologists have
agreed with this conclusion and it is accepted here.

The earlier separation of the gorilla from the shared chimp–
hominid lineage is a profound conclusion. Can a hypothesis be
erected to explain when and why the split occurred? Sibley and
Ahlquist suggested a date for the divergence of the gorilla lineage
at about 8 million years ago. Most of the more recent molecular
studies have come up with similar dates of divergence. However,
molecular methods are best at determining relationships and the
pattern of evolutionary splits. They are less reliable in ascertaining
actual times of divergence. Not only are individual molecules subject to statistical variations in rates of change but the choice of calibration dates has to be taken from the geological record in the
first place, and these dates may also be subject to error. For example, Sibley and Ahlquist put the human–chimp split at 5.5 million
years, and if there is reason from the paleontological record to
move this date to an earlier point in time, then the gorilla divergence date (1.44 times the distance between humans and chimps)
will shift accordingly.

EXPLAINING THE SMATTERING OF GORILLAS

If one is interested in investigating the evolution of the gorilla, it
makes sense to start with what we know about the populations
that are still extant. There are today in Africa three generally recognized gorilla subspecies—geographically defined populations
of animals distinguishable on the basis of their biological characteristics. Subspecies are designated by a third biological name fol-

lowing the species. *Gorilla gorilla gorilla* is the western lowland gorilla, living in dense forests extending from Cameroon and Gabon through to the Central African Republic, and Congo. It was the first gorilla to be encountered on the West African coast by Europeans and to be named by scientists. The western lowland gorilla is found in eastern Zaire and is named *Gorilla gorilla graueri.* And the most famous of the gorilla subspecies, *Gorilla gorilla beringei,* the mountain gorilla, lives in extreme eastern Zaire and tiny montane corners of Rwanda and Uganda. Although their ranges have shrunk over the past century, it seems likely that there were no other gorilla subspecies in the wild of which we are unaware and which disappeared due to human depredation.

Evolutionary biologists consider that subspecies of biological species are defined by both their anatomy and geography. In the case of the gorilla, the anatomical distinctions among the three subspecies are minor, but their ranges are of great significance in understanding their evolution. Since we know that gorillas are denizens of dense forest, our initial expectation then might be that each subspecies would be found located in an island of forest characterized by ancient endemic trees, insects, and other animals. Biologists have discovered three of these so-called forest refuges in Africa, created when savanna grasslands and sparse woodlands spread out and encircled the forests during past climatic drying periods. There are also three gorilla subspecies, but they do not match up to the refuges. Two gorilla subspecies, *graueri* and *beringei,* are found in one—the most ancient of the refuges, the Central Forest Refuge, in eastern Zaire, Uganda, and Rwanda. The third subspecies, *gorilla,* is found in the Cameroon–Gabon Refuge. And the third forest refuge, the Upper Guinea Refuge, lacks gorillas altogether. A more complicated hypothesis is needed to explain the distribution.

The Upper Guinea Refuge is on the extreme western side of Africa and is home to such forest endemics as the pygmy hippopotamus and the mandrill, both of which can tolerate more open wooded conditions as well. The Upper Guinea Refuge is dense forest today but it is separated by a strip of savanna known as the Dahomey Gap, cutting it off from forests to the east. If climates deteriorated and the forests in the Upper Guinea Refuge shrank, not only would forest-adapted species have to tough it out in what trees remained but they also would have no eastward escape route to more extensive forests. Gorilla ancestors would have had a hard time surviving in areas where forests had thinned out to the point that they could no longer find food.

There is one characteristic that both the Cameroon–Gabon and the Central Forest Refuges share. They both have mountains. This fact is significant for the evolution of the gorilla because during phases of climatic deterioration forests extend up mountain slopes. It is here that rainfall is greatest and temperatures (and thus evaporation) are lowest. Mount Cameroon was probably the center of gorilla populations in the west, and the uplifted Virunga Volcanoes and Western Rift Valley mountains (Kahuzi, Biega, and Tchiabirumu) in the east were home to the mountain and eastern lowland gorillas, respectively.

From these data we may conclude that gorillas are and were limited to dense tropical forest, and that their evolutionary centers were tied to montane refuges of this vegetational zone. They were not able to disperse across open savanna or savanna woodland habitats. But the fact that all three gorilla subspecies seem to be closely related would suggest that during periods of greater rainfall and more extensive forests, genetic contact between gorilla populations was reestablished. It is impossible to be sure from these data what may have been the original site of evolutionary divergence for the gorilla, but one evolutionary rule of thumb holds that the area with the greatest diversity of living subspecies is the most likely place of origin of the species. With only a sample of three subspecies, applying this rule to the gorilla does not inspire confidence, but the weak indication is that the Central Forest Refuge is the more likely area of its origin.

Taking what we know from the biomolecular record—a separate gorilla lineage by perhaps around 8 million years ago—and what we know from the modern biogeography of the gorilla—that the species is distributed in accordance with dense tropical forest and montane areas—we can conclude that its origin could be related to ecological or geographic changes occurring at about this time and affecting the distribution of African forests. What can the paleoclimatic record provide us that can help to frame more precisely our hypothesis of gorilla evolution?

THE WORLD OF THE AFRICAN MIDDLE MIOCENE

The middle Miocene was that time extending from about 8 to 15 million years ago. A dramatic and long-term trend toward globally cooler temperatures started at the beginning of this period, 15 to 16 million years ago (Figure 3). On land, temperatures were generally lower and precipitation levels dropped. Sparsely wooded grasslands made their first appearances in Africa.

Eastern Africa followed the trend in increasing the general openness of vegetation. The fauna from Fort Ternan, Kenya, dated to about 13 million years ago, indicates that the vegetation had become more savanna-like. This first antelope-dominated fauna shows that grazing animals had become an important part of the overall ecology of the time. But as we have seen, the hominoid that lived at Fort Ternan, *Kenyapithecus*, could not have been the ancestor of the hominids, chimps and gorillas. Although it was the right size (small) to have been an appropriate ancestor, it was a modified palm-down-walking quadruped that had, so far as we know, none of the derived anatomical specializations of the African great apes and humans. At about 13 million years ago, it was already committed to a woodland–savanna environment too sparsely vegetated for either gorillas or chimps, and thus from a paleoecological standpoint it is unlikely to have been a common ancestor. This is an appropriate environment to hypothesize for an ancestral hominid, but *Kenyapithecus* is far too early to represent a separate hominid lineage, which the biomolecular evidence puts at between 4 and 8 million years ago.

A recent high-tech debate has raged over whether Fort Ternan records the first appearance of African grasslands or not. Gregory Retallack of the University of Oregon reported that he had found both fossil grasses and ancient preserved grassland soils at Fort Ternan. He concluded that by 13 million years ago grasses had become a significant part of the African ecosystem, a conclusion in agreement with expectations from the global oxygen isotope record. But Thure Cerling at the University of Utah, whose work on carbon isotopes throughout this period showed that grasses did not become a major part of terrestrial ecosystems until about 7 million years ago, contested Retallack's conclusions. Cerling's findings relate to the fact that savanna grasses today utilize a different carbon metabolic pathway, one that incorporates a heavier form of carbon known as C-4. Most other plants use lighter C-3 carbon in their cellular activities. Cerling found that the carbon isotopes in the carbonates of fossil soils that he tested in Pakistan and Kenya did not become heavier until about 7 million years ago, thus contradicting Retallack's conclusions. Other researchers entered the fray and measured the carbon isotope values at Fort Ternan itself. They found that indeed there was a shift to heavier carbon values 15 million years ago, corresponding to the global cooling trend. A lesson was learned. Carbon isotope values of fossil soil carbonates record regional climatic conditions and cannot be extrapolated too far afield.

Was *Kenyapithecus* an Ancestor of the Gorilla?

Kenyapithecus has been a controversial ape since Louis Leakey gave birth to it in 1963 at Fort Ternan. Leakey soon afterward wrote an article on Africa as the "cradle of mankind," with *Kenyapithecus* in the role as bouncing baby boy. This hominoid, Leakey thought, showed the relatively small canines, the short-

ened muzzle, and the parabolically shaped tooth row of the hominids. Unfortunately, the fossils were small and scrappy and did
not include any of the critical parts of the face or skull. Other paleoanthropologists remained interested but were cautious in their
acceptance of Leakey's claims. Leakey tried, in vain, to convince
his colleagues that a large stone found at Fort Ternan showed
damage resulting from hominid-like tool use by *Kenyapithecus.*
Virtually no one believed this suggestion, but it was at least
widely cited. Then Leakey came up with the idea that *Kenyapithecus* and most of the other fossil animals at Fort Ternan had
been asphyxiated by poisonous gases vented near the site, and
this one was not even cited. Most paleoanthropologists, who for
the most part had not studied the original fossils, adopted a wait-
and-see attitude until what was hyperbole and what seemed scientifically defensible about *Kenyapithecus* was sorted out. Later
researchers, as discussed in Chapter 1, allied *Kenyapithecus* with
the Asian *Ramapithecus,* when the Early Divergence Hypothesis
held sway, and both apes were thought to have something to do
directly with hominid evolution. It was pointed out that *Kenyapithecus* has molar teeth with thick enamel, like hominids and unlike modern African apes.

Discussing *Kenyapithecus* in the context of gorilla evolution
initially seems inappropriate. First of all, it lived in the savanna
woodlands, not a gorilla environment; second, it has always
been associated with hominid, not African ape, evolutionary origins; and third, it is too old, if the biomolecular analyses are
anywhere near correct, to have anything to do with gorilla–
chimp–hominid divergence. But *Kenyapithecus* and its ape relatives in the Kenyan middle to late Miocene are the closest candidates for fossil ancestors of the living apes and hominids that we
have, so it is necessary to examine them closely before we go on
to other hypotheses.

Kenyapithecus, and its middle-to-late Miocene relatives in East
Africa—newly discovered hominoids named *Turkanapithecus,
Afropithecus,* and a couple more not yet christened—represent an

evolutionary adaptation to woodlands and perhaps even woodland savannas. These hominoids remain poorly known. One of them, represented by only a mandible from the site of Samburu Hills, Kenya, and dated to about 9 million years ago, is large and has been hypothesized as a possible gorilla ancestor by Ishida. Could *Kenyapithecus* have been the ancestor of the Samburu hominoid, and could this hominoid have been ancestral in turn to the gorilla?

To answer this question we must look at what in *Kenyapithecus* might preclude this scenario. First are its teeth. The molars are flat and have thick enamel, unlike gorillas and chimps, which have thin enamel on somewhat cusped molars. Thick enamel is generally considered a hominid trait, but studies by Lawrence Martin, of the State University of New York at Stony Brook, determined that it was probably a characteristic of the common hominid–great ape ancestor. Thin enamel, in the days of the Early Divergence Hypothesis, was thought to have been a primitive feature in the living chimp and gorilla, retained from early Miocene proconsulid apes, which also have thin molar enamel. But Martin's studies showed that the modern African great apes have secondarily thinned enamel that develops slowly. This type of thin enamel is different from early Miocene apes and from living gibbons, which have thin enamel that develops rapidly. Martin hypothesized that the ancestor of all the hominids and all the great apes, including the orangutan in Asia, had an ancestor with thick (and rapidly developing) enamel. Thus, *Kenyapithecus* was not so bad a potential ancestor after all, and several paleoanthropologists, including myself in 1983, suggested that it would be an appropriate ancestor for the hominids and African great apes.

Other aspects of *Kenyapithecus,* such as its small size and relatively small canine teeth, were not major problems. It had long been thought that the gorilla had evolved from a smaller hominoid closer in size to the chimp. But then in 1993 paleoanthropologists Brenda Benefit and Monte McCrossin at Southern Illinois University discovered at the middle Miocene site of Maboko, Kenya, the

upper-arm bone, the humerus, of *Kenyapithecus*. This bone had
anatomy that made it clear that its possessor had possessed more
of a quadrupedal monkeylike style of movement than that of a
true ape, which can climb hand-over-hand and hang from its uplifted arm. For most paleoanthropologists, this discovery has
proved fatal to any claims that *Kenyapithecus* might have retained
for its inclusion in the evolutionary lineup of the hominids or the
living apes.

The large Samburu hominoid, dating some 2 or 3 million years
later than *Kenyapithecus*, has no postcranial bones associated with
it, so we do not know how it moved around. There is, however,
nothing in its anatomy other than its large size that specifically relates it to the gorilla, and we already know that large body size can
be attained quite independently in primate evolution. The most
convincing argument against the inclusion of the Samburu hominoid in the ancestry of the gorilla comes from its context. It is
found at a site with a savanna-woodland fauna—antelopes, giraf-
fids, and other open-country animals—very unlike the environment in which gorillas live today. And the site is some 1,000
kilometers to the east of the nearest living gorilla. Unless we hypothesize that the ancestor of the gorilla moved from a densely
forested environment into a savanna woodland environment and
then back into the same densely forested environment, where gorillas are found today, then we have a rather unlikely scenario for
including the Samburu hominoid in gorilla ancestry.

Kenyapithecus and its late Miocene East African relatives seem
to have evolved independently of the African great apes or hominids. Had wooded savanna and woodland conditions stayed the
same in eastern Africa, these species would probably still be
around today (and we probably would not be). But conditions
over the next several million years continued to become drier, and
the vegetation became sparser. At the time that *Kenyapithecus* was
alive in eastern Africa, our ancestors, which at that time we shared
with the gorilla and chimpanzee, were living in the dense tropical
forests many kilometers to the west.

GORILLA ORIGINS IN THE MIDDLE
AND LATE MIOCENE FORESTS

What of the other parts of Africa—the forests—where we suspect that the common ancestor of the African great apes and hominids lived? Unfortunately, much remains unknown of these mysterious places, but in recent years some progress has been made in understanding what was going on there in the middle and late Miocene.

It used to be thought that the paleontology of the forested regions was unknowable, and this is still a widely held belief. Because forests have acidic soils, bones, which are alkaline, tend to disintegrate. While this is true, it ignores the fact that bones in any environment will be broken up and weathered down by the elements if they are not buried. It is the rate of burial of bones in any environment that determines whether they will be preserved, not necessarily the chemistry of the soil on which they initially rest. Many of the sites that preserve remains of primates, from the Paleocene through the early Miocene, are in fact from forested environments. They just happened to be in geological areas that were subsiding rapidly and thus burying their sediments and bones.

The African Rift Valley is (and was) an excellent place for rapid burial of sediments, including bones that eventually become fossils. Most paleoanthropological attention has been focused on the Eastern Rift Valley, in Tanzania, Kenya, and Ethiopia. But the Western Rift Valley, running along the Zaire–Uganda border and southward past Rwanda and Burundi into Lake Tanganyika, also preserves ancient buried sediments along its flanks. The Western Rift Valley runs right through the eastern portion of the Central Forest Refuge, and so it can serve as a focal point for testing anthropogenic hypotheses relating to the evolution of the gorilla and chimpanzee, as well as perhaps the earliest hominids.

The oldest deposits in the Western Rift are in the region of the Sinda and Mohari Rivers, small streams in eastern Zaire that drain the east-facing wall of the Western Rift Valley and flow into the Nile via its direct tributary, the Semliki River. The Sinda and Mohari erode sedimentary deposits that cling precariously to the extreme margin of the Western Rift Valley. These sediments rest on ancient Precambrian basement rock and bear mute testimony to the opening up of this great gash in the earth's crust. As the earth sank between two giant linear faults demarcating the Western Rift Valley, flowing water in the bottom of the rift laid down silts, sands, and clays. Bones of the animal species alive at the time were deposited as well. Millions of years later, when upfaulting brought the buried deposits back to the surface, erosion exposed them and their fossils to the prying eyes of paleontologists.

The fossils are there but they are very rare and difficult to find. The terrain is rugged and no roads have been built into the rift. Lengthy foot safaris à la Henry Morton Stanley (who named the Semliki River in 1878) are the only way to reach the sites. Unlike most of the Eastern Rift, the Western Rift receives a far greater amount of rain and its vegetation cover is correspondingly dense. Fossils that have been exposed by erosion are usually obscured from view by the overgrown vegetation. Despite these obstacles, research by our project and by Belgian, British, French, and Japanese colleagues on both sides of the Zaire–Uganda border has given us a basis for outlining a general framework for what was happening in Central Africa during the Miocene.

It is clear that this part of Africa retained a very conservative, forest-adapted fauna. Miocene turtle fossils are present here that, like the lemurs, are extinct on the African mainland but persist in Madagascar. The okapi, a short-necked giraffe reminiscent of Oligocene giraffes, is documented here and has persisted in the Central Forest Refuge of Zaire up to the present day. Forest proboscideans, mastodonts with low-crowned molars very much unlike modern elephants, lived alongside pygmy hippos. A totally extinct group of hippo- and pig-like animals, the anthracotheres, are also known from Sinda–Mohari. These were thought at first to be indicative of an early Miocene age, since these animals went extinct in eastern Africa after about 18 million years ago, but they apparently held on in Central and North Africa until much later.

Not only have new potassium-argon dates shown that the Western Rift Valley cannot be more than about 8 million years old, but other species found associated with the anthracotheres are clearly much younger.

In the fossil assemblage from Sinda–Mohari is a lone tooth of a primate—surprisingly, a monkey. Monkeys are rare and little-known primates in the early Miocene sites of East Africa, compared to the diversity of proconsulid apes. They became a more prominent part of middle and late Miocene faunas as ape diversity waned. There is not a single fossil of an ape from Sinda–Mohari, a further indication that these still poorly dated sites are probably middle to late Miocene in age.

How can we be sure that fossil apes were present in the Sinda–Mohari fauna if none have been found? Biogeography is the strongest argument. Near the site where my field crew discovered a mastodont molar, I caught my first glimpse of a free-ranging chimp early one morning. In spite of the relatively dry conditions of today, chimpanzees still live in the area. It is likely that in the past, when conditions were much more forested, apes also lived here. The lone fossil of a possible gorilla, dating to a much later time period, about 3.5 million years ago, is found to the east of here in Uganda. If upheld, this discovery indicates that during the Pliocene epoch gorillas extended from the Zairean rain forests through this region. And finally, Sinda–Mohari is contiguous to the highest elevations in Central Africa, the Ruwenzori Massif and related mountains along the western rim of the Western Rift Valley. The Ruwenzoris, which include the third highest mountain in Africa (after Kilimanjaro and Mt. Kenya), retain ancient montane-adapted vegetation millions of years old. They were natural refuges for the African forests that retreated upslope during the relatively dry periods of climatic history.

The Evolutionary Divergence of the Gorilla

If middle to late Miocene hominoids were present in the Sinda–Mohari fauna, they are likely the candidates that we are seeking for the common ancestor for the African apes and hominids. If the first evolutionary split that we expect was the gorilla/chimp–human split, what would have been the cause of this event?

The first and most obvious possibility to examine is the formation of the Western African Rift Valley. Molecular data show that the split of the gorilla occurred at about 8 million years ago, and this is also the time of the formation of the Western Rift. Could the rift have separated a population of hominoids into protogorillas on the west and protochimp–hominids on the east? The problem comes when the biogeography of the chimp and the gorilla are compared. The range of the chimp completely overlaps the range of the gorilla. Chimps do extend slightly to the east of the Western Rift Valley, and Jane Goodall's chimps at Gombe Stream Reserve are some of the most easterly chimps. But chimps also range far to the west in Africa, far past gorilla distribution. There is thus not a clear division between gorilla and chimp distributions today or in the recent past. This model also does not explain the peculiarly large body size and diet of the gorilla. The Rift Valley hypothesis is discussed in more detail in the next chapter in reference to the evolution of the chimpanzee and the chimp–hominid split.

Another hypothesis, and one that appears to be somewhat more plausible based on current evidence, is that gorillas first differentiated to fill montane forest niches that retreated up mountains in tropical Africa as climates deteriorated and became more arid. The large body size of the gorilla evolved in response to colder temperatures on high mountains, and gorilla diet became specialized on forest trees and succulents. The basic split then among the living African hominoids occurred between high-altitude, dense-forest, large-bodied hominoids ancestral to the gorilla, and lowland, forest-fringe, small-bodied hominoids ancestral to chimpanzees and hominids.

This hypothesis also has a problem. The primary obstacle to its acceptance is the dating of the first contraction of African forests in the middle Miocene, at about 15 million years ago. This is too early for the divergence of the gorilla according to the molecular data. Even if the critical point of forest contraction for cutting genetic connections between protogorilla and protochimp–hominid populations was not reached until the end of the period of global cooling in the middle Miocene, at some 12 million years ago, this is still 4 million years earlier than predicted by the molecular clock. Nevertheless, this hypothesis best explains both the biogeography of modern ape distributions in Africa and the adaptations of the extant African great apes. We will also have occasion to examine the molecular dates in the next chapter.

TESTING THE HYPOTHESIS

Fossils dating to the middle Miocene that are uniquely related to gorilla and to chimp–hominid ancestry are predicted by this hypothesis. Both highland and lowland forest sites in Central and West Africa are necessary to test this hypothesis. Sinda–Mohari is clearly a major target, and despite its low altitude today, it is difficult to say at what altitude its fossils were deposited. One day, probably in the next century, when a foundation or governmental funding agency decides to accord this basic evolutionary question a high enough funding priority, an adequately equipped multidisciplinary expedition to this remote area will provide answers. Other sites undoubtedly exist in areas of past geological subsidence in the central and western parts of Africa. Remote sensing, detailed reconnaissance in forested areas, and new methods of extraction of small fossils will surely net significant scientific rewards. Continuing research in eastern and northern African sites will also help to flesh out the paleobiogeography of hominoids in the middle to late Miocene. Finally, molecular analyses need to be undertaken that investigate rates of change appropriate to the scale of events leading to the relatively closely spaced evolutionary splits between gorilla and chimp–hominid lineages.

We may hypothesize the common gorilla/chimp–human ancestor as the last true obligate denizen of African lowland forest. It was small-bodied and likely not a knuckle-walker, which is an adaptation related to increased body size in a long-armed ape. This ancestor may or may not have had thick molar enamel. As a forest inhabitant it would not necessarily have had to chew tough food objects. Only further discoveries can answer these questions.

The gorilla adapted through evolution to stay in its ancestral habitat—the forest. It followed that habitat as it moved across the landscape and became restricted to mountains. But the chimp–hominid ancestor adapted to a new habitat—a more open lowland vegetation that replaced dense forest over most of the original range of the common ancestor. It is to a consideration of this animal and its ecological history that we will turn in the next chapter.

3

The African Western Rift Valley Split Human Ancestors Off from Chimps

When we speak colloquially of "animals," as compared to "human beings," we encompass a set of organisms ranging from amoebas to apes. This astounding linguistic anthropocentrism has for centuries obscured the fundamentally close kinship that we share with the chimpanzee, a relationship that comparative anatomy and molecular genetics now establish beyond any doubt. In fact, the chimpanzee is so close to us genetically that if we were to use just this criterion, we would likely classify ourselves and chimps within the same species! However, unlike many of the other animal species that are as closely related genetically, human beings and chimps are readily distinguishable by significant differences in anatomy and behavior. An oft-quoted observation is that although the human–chimp gap may be narrow, it is deep.

The fossil record of chimpanzee ancestors has even less to offer in the way of skulls, bones, and teeth than the fossil record of gorillas. Simply stated, there is none. The evolutionary history of the chimpanzee, however, can be reconstructed in general terms, or at least constrained, by what we now know of the contexts of chimpanzee evolutionary history. Much of this knowledge comes from fossils and the record of global climatic change over the past 10 million years.

DISCOVERING A LONG-LOST COUSIN

Biological anthropologists have known for a good many years that the great apes in general were the closest primate relatives to humans. But the realization that the chimpanzee was *the* closest relative is a very recent discovery. In fact, there are still experts who contest this conclusion. Jeffrey Schwartz of the University of Pittsburgh, for example, contends that the orangutan, the Asian great ape, is humanity's closest living relative in the primate order. And some molecular biologists and paleoanthropologists maintain that there is still insufficient evidence to conclusively state that human and chimpanzee are more closely related than are the gorilla and chimpanzee. The question of orangutan evolutionary relationships with hominids has been discussed in previous chapters. There are now sufficient data to make any hypothesis of close relationship, to the exclusion of the African apes, very implausible. Resolving the trichotomy of gorilla–chimp–hominid divergence is more problematical, but as this chapter will show, the paleoecology of chimpanzee evolution lends definite support to the closeness of the chimpanzee–hominid relationship, to the exclusion of the gorilla.

Within the last two decades there have been major discoveries that have revolutionized our understanding of the chimpanzee. In fact, there are two chimpanzees: the common chimpanzee *(Pan troglodytes)* and the bonobo or pygmy chimp *(Pan paniscus).* The advances in understanding have been in two fields: behavior and genetics.

Focusing on the behavioral similarities between humans and apes has not been an easy task for humans. Since antiquity, apes have occupied a position in popular culture as buffoons and clowns, and only very recently have objective behavioral comparisons based on good-quality ethological field data been possible. Jane Goodall's pioneering long-term fieldwork at Gombe in Tanzania stands out as a major contribution to our knowledge of chimpanzee behavior.

Goodall and her team first observed chimpanzees making and using simple tools—termiting sticks—which they used to extract and eat termites from their nests. Other primatologists, the Boesches in Sierra Leone, have now even documented chimpanzees in the wild using stone tools to crack open and eat palm nuts. The similarity with basic hominid tool-using behavior, and the distinction with the gorilla, which has never been observed using tools in the wild, is striking.

Chimpanzees, like hominids, also regularly eat meat. Chimpanzee males have now been regularly observed cooperating in the pursuit and killing of small game. This "hunting" behavior and the dietary preference for meat are behavioral characteristics never observed in the gorilla.

Sex is also a shared characteristic between humans and chimps. Chimps, like humans, are active sexually. Field studies have shown that female and male chimps use sex as an important bonding mechanism. For bonobos, sex has become, as in humans, an important part of social interaction, serving to cement relationships, appease aggression, and smooth over disagreements. Gorillas, on the other hand, show little interest in sex. They copulate only when absolutely necessary. If we extrapolate to psychological comparisons, gorillas probably spend nearly 100 percent of their day thinking about food. There is a high probability, in contrast, that chimpanzees, like the average college student of sociological surveys, daydream about sex at least once a day.

If the behavior of chimpanzees is similar to humans, their genetic similarities are even closer. DNA studies support the conclusion that the human species and chimp species are between 97 percent and 99 percent genetically identical. When chimp proteins are compared with those of humans, the similarities are also startlingly striking. For example, only one of the four chains of the giant oxygen-carrying molecule in red blood cells, hemoglobin, shows any difference whatsoever between humans and chimps. There is *one* difference in this very long sequence. Similarly, there is only one difference in the sequences of chimp and human myoglobin, an oxygen-carrying protein in the muscles. One of the biggest molecular changes between humans and chimps is seen in transferrin, a protein that assists in iron metabolism. This molecule still shows only eight differences. Thus, the biomolecular data are clear in pointing to the close genetic bond between humans and chimps.

AN ECOLOGICAL COMMON BOND

On the surface, chimpanzee and hominid habitats seem very different. Chimps live in forests, and hominids in almost all cases are known to have lived in savanna habitats. Even modern humans, when they move into forested areas, expend great effort to "clear" them until they approximate an appropriately thinned out "savanna woodland," even if this happens to be in surburban Chicago. But the traditional distinction between human and chimp habitats is a false dichotomy. It has also suffered from a human-oriented myopia.

When we look at the difference between the habitat of the gorilla and that of the chimpanzee we can see the magnitude of the error in ascribing them to the same "forest" habitat. Gorillas live in dense forest that must always have substantial rainfall. Chimps can range into much more open environments, and can live well in what is described in Senegal as "savanna," generally open woodland with clumps of trees interspersed with grassland. It is true that in some areas—for example, eastern Zaire and the Central

African Republic—chimps and gorillas live together in the same habitat. But their range of habitats overlap only in these restricted and usually human-modified environments. The core of the chimpanzees' habitat is lowland forest, frequently with open patches, savanna woodland, and forest fringe. It does not range into the high montane forests characteristic of gorillas and it does not generally extend into the true, open savanna of humans. In a real sense, the habitat of the chimpanzee is intermediate between that of humans and gorillas.

The open character of the chimp's habitat struck me the first time that I encountered a chimpanzee in the wild. I was surveying for fossils in the semiarid floor of the Western Rift Valley of eastern Zaire. It was a brief sighting and he was skulking away. His black outline stood out in stark contrast to the light brown and deep green of the surroundings. I remember the environment well. It was a very isolated part of the Western Rift, far away from roads and villages, a part of Virunga National Park seldom patrolled and never visited. The setting was exquisite, just near an ice-cold waterfall, shown on the map as Kirk Falls. But it was open country, at several kilometers' distance from the dense cover of either the Ituri Forest or the gallery forest of the lower Semliki River. It was inconceivable that gorillas would ever traverse this area. There would be nothing for them to eat, the temperatures would be much too hot for them, and there would be nothing to entice them to ever come here. Until I had actually visited habitats of the living chimps in Africa, such as this one, and compared them with the rain forests inhabited by the gorilla, this fundamental difference in adaptation between the two African great apes had escaped me.

The chimpanzee clearly lives in a habitat much closer to that of humans than does the gorilla. Assembled with the molecular evidence allying chimps and humans, behavioral similarities observed between humans and chimps, and more similar dietary adaptations, this evidence argues for a more recent common ancestry for chimps and hominids than either had with the gorilla.

What can we piece together from the paleontological record that sheds light on this hypothesis?

RECONSTRUCTING THE WANDERINGS OF *PAN:* THE HISTORICAL BIOGEOGRAPHY OF THE CHIMP

In examining the distribution of subspecies of the common chimpanzee in Africa, we encounter a straightforward pattern. Chimps live in Central and West Africa. The farthest east they extend is to the eastern edge of the Western Rift Valley in Uganda and Tanzania. Each of three subspecies of the common chimpanzee live in one of the three forest refuges that we saw in Chapter 2: *Pan troglodytes troglodytes* lives in the Upper Guinea Refuge in extreme western Africa, *Pan troglodytes verus* lives in the Cameroon–Gabon Refuge, and *Pan troglodytes schweinfurthi* lives in the Central Forest Refuge. The common chimp and the gorilla are sympatric—occurring in the same area but with somewhat different habitats—in the more centrally located latter two forest refuges. These are the two areas hypothesized in the last chapter to have been the most likely sites for original gorilla–chimp/hominid evolutionary speciation, when climatic change split up the forest habitat of the common ancestor and the protogorilla population became limited to the montane forests. When the forests recoalesced during warmer and wetter times, as now, the populations, now distinct species, came back into contact.

The story becomes more complicated when we add the distribution of the bonobo, the pygmy chimpanzee, known scientifically as *Pan paniscus.* The bonobo and the common chimp share a long common ancestry, but *Pan paniscus* is a separate species, thereby implying that it split off at an earlier date than any of the subspecies of the common chimp. Molecular estimates of the time of this splitting are around 2 million years ago, and we do not yet have any fossil evidence whatsoever to go on.

The bonobo's range does not overlap with that of either the

gorilla or the subspecies of the chimpanzee, from which it is separated by the wide and deep Zaire River and its upper tributaries.
Only after we examine the context of African evolution and ecology does the extremely unusual nature of this observation, known
to most undergraduate anthropology students, strike us. In fact, it
is a paradox. The bonobo lives in some of the densest and most
impenetrable lowland and swamp forest in Africa, yet it is surrounded not by other forest refuge species but by recent immigrants from the savannas and woodlands that have reinvaded. No
one knows how the bonobo's unique distribution has occurred,
but we can suggest a plausible scenario that will need to be tested
by fossil data in the future.

The bonobo lives in a vast area of dense lowland rain forest
south and west of the great northern-arching curve of the Zaire
(or Congo) River that, strangely, no lowland gorillas or common
chimps share. Neither do many ancient forest-adapted species of
Africa. Zoologists and botanists have discovered that what appears to be unbroken primeval forest in the Zairean Basin south of
the Zaire River is actually *not* a forest refuge of Africa. There is a
greatly impoverished forest-adapted fauna and flora here, and
there are virutally no known endemic species, all having immi-
grated more recently from other contiguous (less forested) areas.
Geologists have discovered that underlying the dense tree cover
are fossilized dunes of desert sands that extended northward from
the Kalahari Desert in southwestern Africa to cover much of the
Zairean Basin. Any obligate forest species would have been wiped
out when the trees died out. Only species with a tolerance for relatively open conditions that could hold on in areas of local stands
of trees would have survived. This is what we surmise must have
occurred with the bonobo.

The bonobo is thus the only living species of hominoid, other
than humans, which lives naturally in a historically nonforested
environment. This is not to say that the bonobo does not live in a
densely forested environment now. It does. But historical biogeography shows that it could not have evolved in one. Where did the

bonobo go when its forests or forest fringes disappeared? How
did it repopulate the forests once the rains returned?

Zoologist Jonathan Kingdon of Oxford University suggested
several patterns for the repopulation of the nonrefuge forests of
the Zairean Basin. He based his conclusions on the biogeography
of living mammals. The bonobo may have retreated to the southeast, to highland areas of southern East Africa, where it rode out
the dry times and returned to more optimal forest habitats when
precipitation increased. This is a pattern of distribution shown by
a savanna woodland rodent known as *Petrodomus,* which repopu-
lated the Zairean Basin from southern Tanzania, skirting the Western Rift Valley. Another possibility is that the bonobo may have
survived in isolated patches of forests near the Atlantic coast. This
seems a little more plausible based on the distance now between
East Africa and the bonobo range. What is certain is that the
bonobo did not cross the Zaire River to the north and northeast,
to extend its range into common chimp territory.

PUTTING THE PIECES TOGETHER:
A MODEL FOR THE CHIMP–HOMINID SPLIT

When we last left our evolutionary narrative at Hypothesis 1, the
gorilla had split off from the common chimp–hominid ancestor
at perhaps some 8 to 12 million years ago. We deduce from modern ecological data as well as paleoecological context that the
chimp–hominid ancestor was a lowland forest and forest-fringe
species whose populations surrounded the montane forest habitats of the gorilla. The hypothetical distribution of this common
ancestor from 8 to 12 million years ago must have been from the
extreme western parts of the tropical African forests bordering
the Atlantic, extending eastward around the great curve of the
Zaire River to the southern Zaire Basin, and finally into eastern
Africa through what is today Tanzania, Uganda, western Kenya,
and the northwestern half of Ethiopia, up to the Eastern Rift

Valley. We must piece together a series of events that can account for the separate origins of the common chimp, the bonobo, and ourselves, the hominids, from this original ancestral population.

The first evolutionary split to examine is the earliest, the chimp–hominid split. In searching for a common denominator in chimp and bonobo biogeographic distributions then, it is clear that they both occur west of the Western Rift Valley in tropical Africa. Looking at the known distribution of the earliest hominid sites, it is clear that they all occur east of the Western Rift Valley. If we make the deduction that it was the formation of the Western Rift Valley that cut off an eastern part of this population to become hominids, how might this have happened?

A casual visitor to either the Eastern Rift Valley in Kenya or the Western Rift Valley in Zaire can observe the changes in habitats as one moves through the topographic relief in these areas. There are marked ecological differences as one moves into and out of the rift valleys. On the road west from Nairobi, Kenya, for example, one descends a dramatic escarpment of the Eastern Rift, leaving the relatively well-watered Kenya Highlands with tea plantations and abundant greenery, to enter a low-lying rift valley floor with a dry, acacia-studded landscape and low-lying lakes. Another 750 kilometers to the west, in the Western Rift of Zaire, one can drive up out of the hot rift valley floor habitats of savanna grassland in Virunga National Park to ascend the Kabasha Escarpment and arrive at the top in a cloudy, cool, and forested environment. Originally a chimp habitat, it is now heavily cultivated.

The rift valleys act as linear demarcation lines of wet and dry habitats. Their down-dropped basins allow rain-carrying clouds to blow over, whereas the uplifted margins of the Rift Valleys stop the rain and sustain denser vegetation. The western margin of the Western Rift is the highest in the African Rift Valley System, stops the most rain, and has the densest vegetation. It is a major biogeographic barrier today and its influence in the past, particularly on our evolution, was profound. It is reasonable to entertain the hy-

pothesis that it was the formation of this extensive east–west barrier to genetic exchange that effected the split between the eastern
hominids and the western chimp/bonobo ancestors.

Dating the formation of the Western Rift has been problematical until recently. New potassium–argon dates in the southern part
of the Western Rift near Bukavu, Zaire, indicate that this portion
of the rift is about 5 million years old. Regional geology indicates
that the rift would have begun in the north and propagated to the
south. Estimates date the first formation of the Western Rift in the
north at around 8 million years ago. The meager fossil evidence
from the Western Rift does not contradict this age estimate, even
though it does not provide any firm suppport.

If we accept that the Western Rift was the formative event in
splitting hominid and chimp ancestors, then 8 million years ago
provides an earliest possible date for the evolutionary origin of
both chimps and hominids. Molecular evidence, as we have seen,
indicates a range of between 5 and 8 million years for this split.
Thus, there is general agreement between geological and molecular analyses, although there is still quite of bit of uncertainty. For
example, did the initial rifting effect geographic changes that
caused the complete separation of the populations early in this
time range or was the split only completed later as the Rift Valley
formed a longer north–south front? Better dating of the geological
formation of the Western Rift Valley and more molecular analyses
will provide more precise ages for these determinations and will
allow us to refine the model.

THE FURTHER EVOLUTION OF THE CHIMP AND THE BONOBO

Between some 2 and 8 million years ago, the common chimp and
bonobo would have shared a common ancestor. This gene pool
would have extended throughout the Zaire Basin, on both sides of
the Zaire River, and into the more northerly forests. What would

have occurred at about 2 million years ago to split the northern and southern populations of chimps into two?

Global climatic patterns, to be discussed in the next chapter, became markedly different beginning first at about 2.8 million years ago. Evolutionary effects of these climatic oscillations are seen in the paleontological record in Africa by at least 2.5 million years ago. Dating to between 2.0 and 2.3 million years ago, there was a significant period of aridity in the Western Rift. At our fossil sites in the Upper Semliki Valley, major extinctions of lake-living molluscs, and the fish that ate them, indicate that Lake Rutanzige (Lake Edward) dried up to only some small pools. The land-living mammals at this time were all savanna open-country species, virtually identical to similar-aged sites in the Eastern Rift. Early hominids lived here and left their stone tools.

Clearly, by 2.0 to 2.3 million years ago, the corridor of forest running eastward around the northern bend of the Zaire River and into the Western Rift Valley, that had connected the northern chimp–gorilla region with the bonobo region of the south, had been cut. The paleoclimatic data showing this change in habitat roughly corresponds with the molecular data of the chimp–bonobo evolutionary split. But as with other molecular evolutionary dates correlated with paleontological dates, they are on the young end of the possible range. It seems very likely, however, that after about 2 million years ago bonobos were restricted to the forest regions south of the Zaire River, and they never regained a broader distribution.

Common chimps occupy a much wider region of tropical Africa than does the bonobo. The three subspecies of *Pan troglodytes* correspond to the three known forest refuges. Clearly these three centers were separated from each other in the past by areas of savanna grassland. These swaths of open vegetation had to have been wide and arid enough to curtail movements of the forest-living chimps. They therefore cut population connections between the three core areas of chimp populations. The dates of these disjunctions had to postdate the evolutionary divergence of the common chimp and bonobo at about 2 million years ago, and probably are due to the increasingly severe climatic fluctuations of the Pleistocene epoch, to be discussed later. These three population centers of chimpanzee evolution could very well be as ancient as the bonobo–common split, but genetic exchange among the three population centers has almost certainly been possible when unbroken forests connected all three areas. These episodic periods of gene exchange probably maintained genetic connections and blurred what could have become species-level differences among the three subspecies of common chimps.

A New Perspective on the Bonobo as Hominid Prototype

The primatological researchers who have studied the behavior and adaptations of the bonobo in its natural environment in central Zaire have not paid an inordinate amount of attention to the paleoecological aspects of the bonobo's evolution. This is understandable as behavioral research on the bonobo is still in its infancy and many facets of the species' ethology remain to be investigated. The fact that the bonobo is the least well known of the living hominoid species belies that it has occupied such a prominent place in hypotheses on hominid origins.

The discoverer of the bonobo, Harold Coolidge, wrote in 1933 that the species "may approach more closely to the common ancestor of chimpanzees and man than does any living chimpanzee." Anthropologist Adrienne Zihlman of the University of California at Santa Cruz and several coauthors took up this thesis and proposed in 1979 that of all the living hominoids the bonobo was the one closest to the anatomy and behavior of the common human–ape progenitor. Many of the anatomical ideas incorporated in Zihlman's work came from a doctoral thesis written several years earlier at the University of Chicago by Douglas Cramer. Cramer had undertaken a detailed study of the bonobo's skull, using the largest collection of these primates, housed in the Royal Museum of Central Africa in Tervuren, Belgium. It was his thesis that the bonobo had a more hominid-like cranial structure, smaller canine teeth, a more globular skull, and a less protruding face. He pointed out that of all the living apes, the bonobo was the most similar in gross anatomy to the human.

Zihlman extended these ideas to include the postcranial skeleton. The upper limbs were relatively shorter and the lower limbs were relatively longer in the bonobo than in the common chimp, and in these characteristics the bonobo was also more similar to humans. Although sometimes called the "pygmy chimp," the bonobo is actually not much smaller than the average common chimp; it just has a more lightly built upper body. Bonobos walk bipedally more frequently than common chimps, and males and females tend to copulate facing each other. There is less difference in size and body shape between male and female bonobos than between the sexes of the other great apes. All these characteristics ally the bonobo with humans.

Zihlman and her colleagues stopped short of suggesting why the bonobo may have shared some or all of these characteristics with hominids. A paleontologist, Steven Johnson, suggested in an article in *Current Anthropology* that the bonobo evolved in a manner analogous to an island relict, isolated in a patch of forest shrunken by the vegetational vicissitudes of the Pleistocene. He focused on the idea of the bonobo as a "pygmy," and made a comparison to such cases of evolutionary dwarfism as the tiny ice-age elephants stranded on Mediterranean islands. Perhaps, he suggested, the small size of the bonobo was the primary factor that made it appear so humanlike, compared to either the common chimp or the gorilla.

Despite the fact that the bonobo may have been limited to restricted forest patches in its evolutionary history, it is inaccurate to describe it as a "dwarf" species. Its average body weight is 35 to 40 kilograms, only slightly lower than the common chimp. Actually, female bonobos and female chimps are the same body weight, less than 30 kilograms. It is the larger body size of the largest males that accounts for the overall body size difference between common chimps and bonobos. The bonobo's lack of "sexual dimorphism," estimated at less than 25 percent, is yet another aspect of its anatomy that is comparable with humans. But if its relictual evolutionary history leading to "island dwarfism" does not explain its humanlike anatomy, what does?

An ecological perspective provides some possible answers. There may well be several functional connections between the fact that the bonobo does not live in an African forest refuge and the fact that it shares many similarities with humans, who also originate from nonforested areas of Africa.

The bonobo shows more bipedality, a human characteristic. Bipedality is an adaptation for traversing long tracts of territory and is seen in hominids more than in any other primate. If the bonobo in the past had to cross open patches of vegetation to reach the next stand of trees, bipedality may well have been an important part of its behavior. Thus, although bonobos are still committed knuckle-walkers, their locomotor adaptation may include a greater component of bipedalism because natural selection favored those populations able to traverse habitats more open that any encountered by their cousins north of the Zaire. The common chimp, on the other hand, became more committed to a strictly quadrupedal knuckle-walking mode of locomotion, in parallel to the gorilla. Knuckle-walking serves well in traversing forested substrates—both terrestrial forest floors and large, subhorizontal arboreal branches—but it is a poor, energy-wasting way to cross open terrain. The common chimp is correspondingly more heavily built in its upper body, with heavier arm and shoulder musculature, than the bonobo. Indeed, this difference in body build accounts in the main for the appellation "pygmy" being applied to the bonobo.

Even more striking than the physical differences between common chimps and bonobos are the differences in social behavior of the two species, recently described by several teams of primatologists working in Zaire. Bonobos seem to have a different and more

complex social fabric than the common chimp, a factor accounting perhaps for more stable intragroup relations and the observed
generally larger size of their social groups. Bonobo group organization seems to revolve around a stable core of females, but strong
"friendships" between females and males, usually involving sex,
are very important in interindividual dynamics. Sex plays a much
more important social role in bonobos than in common chimps,
and females copulate and show genital swelling through significantly longer portions of their menstrual cycles than do common
chimps. Sex is used socially by female and male bonobos to form
alliances and for appeasement ("making up"). Sex in common
chimps, in contrast, is restricted to the females' period of estrus
(or "heat"), which tends to be synchronized among the various females in a common chimp group. Bonobo females do not show
this synchronization. In this regard they may be similar to human
females, who generally do not show intragroup synchronization of
menstrual cycles and show no external signs of estrus (ovulation)
at all.

Male common chimps bond strongly to other males and have
been observed not only to "hunt" cooperatively but to carry out
chimp "wars" with other groups, carry off females, and kill and
eat other chimps, including infants. This degree of male solidarity
has not been observed in bonobos, where all-male groups, fairly
widespread in the common chimp, are very rarely or never observed. Female common chimps do not seem to form the strong
affiliations with other females observed in bonobos. Bonobos
seem to have a more "egalitarian" social structure as regards males
and females, which may or may not be characteristic of humans,
depending on your point of view. Finally, some research indicates
that bonobos have superior communication skills to common
chimps. Considering the more complex interindividual social interactions of the bonobo, this adaptational difference would not
be surprising.

Two conclusions emerge from this discussion. First, there are
both physical and behavioral differences between common chimps

and bonobos. Second, where the bonobo differs from the common chimp, it resembles humans. There is a clear ecological and adaptive reason for the bonobo's lightness of upper body build that follows from the scenario of bonobo evolution painted here. Climatically induced opening up of vegetation in the bonobos' habitat placed a natural selection premium on the ability to traverse open terrain effectively. A greater degree of bipedalism in the behavioral repertoire of the bonobo thus resulted.

Explaining the behavioral similarities between bonobos and humans based on what we know or suspect of the bonobo's evolution is more difficult. Primatologists have speculated that the larger group sizes of bonobos are a result of a more predictably abundant forest food and resource supply. While there is nothing in the current forest ecology of the bonobo to contradict this notion, the ecological history of *Pan paniscus,* as painted here, would certainly lead to a very different conclusion. It is much more likely that the greater degree of sociality in the bonobo is a result of strong natural selection for intragroup cooperation under situations of very scarce resources. Bonobo groups survived by cooperating in locating and sharing environmental resources. Large groups with many cooperating individuals would have been at a clear competitive advantage because wide areas could be effectively surveyed for resources. Forming affiliations, "friendships," and sexual liaisons would have been very important in such a cooperative adaptation. It is unlikely that such elaborate social mechanisms would have evolved simply in response to environmental abundance of food resources. Cooperation thus evolved in the bonobo as a consequence of ecological scarcity, not abundance. The scarcity was brought on by climatic change.

It is instead the common chimp, a denizen of ancient forest refuges, that exhibits an adaptation in social behavior based on a predictable resource base. The overall area to which this predictability extended varied through time, contracting at points in ecological history, but the common chimp was able to hold on in isolated but internally unchanging forest/forest fringe refuges. Its social behavior stayed much the same as the sexually dimorphic apes from which it had sprung. Male and female common chimps continue to exhibit seasonal reproductive behavior as well as more individualistic foraging, food-getting, and food-sharing behavior. Only males cooperate in group food-getting—hunting—and even this activity is rare.

The bonobo is thus a most instructive case for hominid evolution. The species is no closer genetically to humans than is the common chimp, but its ecological history in some important ways parallels that of our direct ancestors. It is therefore an *analogy*—a similar evolutionary product not caused by common ancestry but by similar forces of natural selection—to the human condition. When the earliest currently recognized hominid, dating to 4.2 million years ago, was discovered in northern Ethiopia in 1994, it was not surprising that its teeth bore a strikingly close resemblance in both shape and dimensions to those of the living bonobo.

ADRIAN KORTLANDT'S VIEWS OF CHIMPANZEE ORIGINS

Dutch primatologist Adrian Kortlandt published in 1972 a book entitled *New Perspectives on Ape and Human Origins* that set forth a biogeographic argument that also posited a formative role for the Western Rift Valley in separating chimp and hominid ranges. Kortlandt's research was primarily in the realm of behavioral adaptation of the common chimp. He pioneered the use of recording primate behavior in the wild on film. His most famous piece of research was a naturalistic experiment, recorded on film, that he carried out near Beni, eastern Zaire. Kortlandt planted a stuffed leopard near the path of a chimpanzee group, which reacted with predictable alarm when they first saw it. The males of the group attacked the leopard with branches and sticks, at first tentatively, and then with increasing fury. Ultimately the leopard was dismembered, with its stuffing falling out—totally "killed." The chimps then lost interest. It was an impressive display not only of chimpanzee tool use but also of the ability of this cousin of the passive gorilla to defend itself, much like a human.

As a zoologist, Kortlandt pointed out that the pattern of African rivers, rift valleys, and vegetation zones would have affected the evolution of the chimpanzee. He emphasized the importance of the Western Rift Valley at a time when virtually all other anthropologists were focused on the Eastern Rift. The Western Rift does indeed appear to be pivotal in establishing separate protochimp and protohominid populations. Although Kortlandt suggested a much more major role for the Nile River in effecting this separation than is clearly warranted by more recent data, the importance of the Zaire River in delimiting the bonobo's range from those of the common chimp and gorilla remains valid. Perhaps the most important aspect of Kortlandt's contribution was that it established a biological framework for looking at later African hominoid evolution that was not canted entirely to the human side.

Paleoanthropologist Yves Coppens followed Kortlandt's lead in suggesting in 1983 that the African Rift Valley system was formative in separating apes from hominids. Coppens, in a play on the Broadway musical, termed his hypothesis the "East Side Story." He emphasized the distribution of fossil hominid sites only in eastern and southern Africa, the lack of fossil apes in any of these sites, and the distributions of the extant African apes exclusively to the west of the Eastern Rift Valley (west of which no fossil hominids were known). He and other workers have implicated climatic change toward more arid conditions as important in these evolutionary events. We will have occasion to revisit Coppens's ideas more fully in Chapter 6.

TESTING THE HYPOTHESIS

Hypothesis 2, which results from the discussion in this chapter, states that the chimpanzee and hominid evolutionary split occurred because the Western Rift Valley formed and separated these two populations. The date of this split was approximately 8 million years ago. Several tests are needed to confirm or falsify the predictions made by this hypothesis.

First, the time of formation of the Western Rift needs to be more firmly dated. If the date of formation can be firmly established at 8 million years ago, then this is the earliest possible date for chimp–hominid divergence. Second, the molecular evidence for chimp–human divergence needs to be retested using alternative molecules and alternative methods. Parasitological molecular evolution may provide a very valuable tool. In this method, the molecular evolution not of the host organism but of the populations of coevolving parasites is tested. Third, paleontological reconnaissance in the regions of earliest chimp–hominid evolutionary divergence has to be expanded. The goal of this research is to recover the fossil remains of the common chimp–hominid ancestor and document the anatomical changes that were attendant to the split. In this day and age of decreasing research budgets and the need to show immediate practical and applied benefits, such a basic research program will be difficult to fund. Nevertheless, it will be necessary for a full test of this model. Finally, the ecological parameters that went along with this evolutionary divergence need to be investigated by the many and varied tools that paleoanthropologists now have at their disposal. A fit needs to be established between the observed anatomical changes and the changes in habitat that occurred at this time.

Hypothesis 2 has much to recommend it and it is impossible at this point to disprove it. But there is another series of hypotheses that is also compatible with the fossil record, paleoclimatic history, the known distributions of living African apes, and the data of molecular evolution. These hypotheses are set later in time and focus on another part of Africa, one not traditionally associated with the well-publicized discoveries of paleoanthropology. It is to North Africa and the Mediterranean Sea that we will next turn in investigating the timing and ecology of the last ape–human evolutionary split.

4

The Developing Sahara Split Hominids Off from Chimps

In 1969 a group of oceanographic scientists on the *Glomar Challenger* found themselves anchored over one of the deepest basins of the Mediterranean Sea. The *Glomar Challenger* was a research vessel and this was Leg 13 of the Deep Sea Drilling Project, an international effort to understand the geology and biology of the world's oceans. Surrounded by the deep blue of the Mediterranean, the scientists had no idea that they were about to make a discovery that would revolutionize our understanding of the largest desert in the world and pave the way for a new theory of human origins.

One of the shipboard geologists looked at the readout sheet that was coming out of the gravity reflectance machine. When it finished printing he tore it off and spread it out on the big table in the chart room. As he looked over the wavy lines of the reflected sonar waves, he frowned. Some of the others sauntered over to the table, but when they glanced at the printout they fell silent. For a fleeting moment there was a spark of fear that spread through the group—*we don't know what's going on*. Then they recovered. Their faces assumed a sort of vacant deliberateness. They were confused but thinking hard.

Reflectance data give information on the geological strata that underlie the sea bottom. Because the ship was over a deep basin, the scientists expected to see some big faults and bedrock layers plunging downward under the sea bottom. Instead they were getting a very puzzling readout of a widespread, absolutely horizontal layer that was well above bedrock. "What the hell is that?" one of them finally asked out loud. Since none of them had the faintest idea, they did what scientists who find themselves confronted with an unexpected phenomenon always do—they named it. After that, everyone felt a little better. "Reflector Level M" at least was pinned down and labeled while they took a closer look.

But Reflector Level M refused to stay pinned down. It haunted this cruise like the albatross of the *Ancient Mariner*, hovering over the ship and causing general consternation among the scientific crew. It should not be there, and yet there it was. This ship was supposed to be studying the structure of the deep-sea bottom and Reflector Level M was not allowing anyone to concentrate on the deeper sediments.

The *Glomar Challenger* was also outfitted for drilling into the seafloor and extracting long cylinders of sediments—cores—hundreds of feet down. As the long cylinders of muddy but firmly consolidated sediment were laboriously hauled up on deck and laid out, the scientists honed in on what might be sending the strong reflectance signal back from "Level M." They correlated the length of the core with the depth of the reflectance level, and they found the sediment. But rather than laying the albatross to rest, the answer raised even more questions. The sediment was salt! Specifically, it was gypsum (potassium chloride) and halite (sodium chloride, or common table salt).

A STARTLING DEDUCTION

The scientists on the *Glomar Challenger* knew that there was only one way to get massive deposits of salt at the bottom of a water basin. The water must have evaporated, shrinking into smaller and smaller lakes or pools, concentrating the dissolved salts. Today some East African Rift Valley saline lakes and the Great Salt Lake in Utah are examples of the phenomenon of salt deposition in relatively small water bodies subject to periodic drying. But could they really be analogous to the Mediterranean—a *sea?* The Mediterranean holds over a million square miles of water and in some places is well over 2 miles deep!

By 1970 the chief scientists of Leg 13 of the Deep Sea Drilling Project were ready with an interpretation. Kenneth Hsü of the Eigenische Technische Hochschule in Switzerland and Bill Ryan of Lamont Doherty Earth Observatory in New York published a paper that concluded that Reflector Level M was indeed an indication of the unthinkable. They deduced that the entire Mediterranean Basin had dried up and much of it had been low-lying dry land. Only in some extremely concentrated "lakes" in the deepest parts of the basin had water remained, and here the salts of Level M had been laid down. Had people been around at the time they could have headed south by foot from the Riviera and walked to Morocco! But people had not been around at the time. The dating of the deep-sea core showed that Level M was deposited between 5 and 6 million years ago, between the end of the Miocene epoch and the beginning of the Pliocene. A further deduction was that at the end of the Miocene, when the Mediterranean Basin refilled with water from the Atlantic, there would have been at the Strait of Gibraltar a saline waterfall one thousand times larger in volume than Niagara—a tourist attraction of stupendous proportions just 5 million years ahead of its time.

Perhaps Hsü and Ryan expected to create controversy with their revolutionary idea. Instead, most earth scientists began to see that their data also made sense in light of the new discovery.

Geologists saw a correlation with gypsum-rich deposits on land that had been first studied in Messina, Sicily. These sediments had lent their name to a whole geological stage—the Messinian—and with the new dates from the deep-sea core, the terrestrial deposits could now be better understood. Paleontologists who studied invertebrates also welcomed the new discovery. They saw a tremendous wave of extinctions at the end of the Miocene, and this now made sense if most of the underwater habitats of the shelled creatures that they studied had disappeared. The sudden appearance of new species of snails, clams, and other invertebrates in the Mediterranean with close relatives in the Atlantic at the beginning of the Pliocene made sense because they would have been swept in over the massive Gibralter waterfall.

Scientists began to speak of the *Messinian Event,* a term connoting the drying out of the Mediterranean Basin and the deposition of very saline sediments during the period of time then thought of as beginning at about 6 million years ago and extending up to 5 million years ago. Paleontologists who studied the past marine faunas and floras that lived in the waters of the Mediterranean of course saw a great degree of change from before to after the Messinian Event. But the scientists who had discovered the Messinian Event and who had become interested in its broader ramifications began to ask the question of whether this "event" in the earth's history might have had more general effects on global climate and therefore on life on land. Five million years ago is also the age of the first fossil hominid in Africa, making a suggested connection between the cataclysmic drying up of the Mediterranean and human origins almost irresistible.

THE FOSSIL JAW FROM LOTHAGAM—THE EARLIEST KNOWN HOMINID

The similarity in timing of the end of the Messinian Event and the appearance of the first hominid fossil in the paleontological record is remarkably close. A half mandible from a site near Lake Baringo in the Eastern Rift Valley of northern Kenya known as Lothagam was discovered by paleontologist Bryan Patterson from Harvard University in 1965. The specimen has the squarish, thick-enameled molar of a hominid, and its mandible is both thick from side to side and high from top to bottom—also hominid traits. Dates on the sediments at Lothagam indicate that the mandible is between 5.0 and 5.5 million years old. Patterson and colleagues published the mandible as a hominid, tentatively referred to the species *Australopithecus africanus,* and suggested that its age was about 5 million years old. In the intervening years many paleoanthropologists, including myself, have studied the specimen in the Kenya National Museum in Nairobi, and agree that it is a hominid. Much critical attention and many geological reanalyses have been undertaken on the dating of the site. There is general consensus that the mandible and the mammalian fauna associated with it are close to the age claimed for them—that is, 5 million years old. After Lothagam, but certainly by 4 million years ago, several sites are now known in eastern Africa with clearly recognizable hominids. Thus, by most accounts, hominids are documented in the fossil record of Africa by 5 million years ago. The name by which the fossil should be known is uncertain, but it is likely another more primitive species of *Australopithecus,* or perhaps a different genus of hominid altogether.

The first appearance in the fossil record of a species or a lineage does not necessarily correspond to the time and place of its actual evolutionary point of origin. We must accept this caveat if we can even entertain the set of hypotheses concerning the origins of gorillas, chimpanzees, and hominids discussed in the preceding chapters, since in these places and time periods the fossil record is woefully inadequate. There are no cosmic rules that say that a species has to originate only where we can find fossils millions of year later. Our Hypotheses 1 and 2 are driven more by the compelling data of molecular genetics, ape biogeography, and African paleoenvironments. But the fossil record is our most unambiguous and clear-cut research method of determining that a particular species was at a particular place at a particular time.

For the purposes of Hypothesis 3, the subject of this chapter, we will accept that the end of the Messinian Event and the beginning of the hominid lineage, as documented in the fossil record, correspond in time. There are still many questions as to whether there is any causal connection between the two. As all scientists know, correlation does not necessarily indicate causation. We can say that correlation is a necessary condition, but not sufficient, to accept that one event in nature causes another. If two events observed in nature or the laboratory seem to be related, further research is needed to ascertain whether there is a chain of events or processes that can explain the correlation. In the case of the Messinian Event and hominid origins, a dramatic geographic and geological event in the Mediterranean Basin culminated at 5 million years ago, and an apparent evolutionary emergence of hominids occurred half a continent away, some 3,000 kilometers to the south, also at 5 million years ago. What kinds of causative influences could be invoked to explain the correlation in timing?

THE MESSINIAN, CLIMATE CHANGE, AND HOMINID ORIGINS

John Van Couvering, an earth scientist at the American Museum of Natural History who early on was interested in the connection between the Messinian and hominid origins, is fond of showing a facetious slide in lectures. It is a cartoon of an apelike creature hopping up and down and shaking its hands while symbols indicating cursing stream from its mouth. The creature, a knuckle-walker, has emerged from a shady forest in the background onto the hot, arid Messinian salt flat at the bottom of the Mediterranean Basin. The hot ground has burned its knuckles, and Van Couvering suggests with a wry smile that this clearly must be the origin of bipedalism, the hallmark of the hominids.

Initial hypotheses connecting the Messinian to hominid origins were not too much more sophisticated, if less amusing, than this scenario. Like Van Couvering's cartoon, they related emergence of the hominids to ecological change. One idea was that as water in the Mediterranean Basin evaporated, there would have been less and less water entering the atmosphere over the Mediterranean. Less rain would have fallen in the surrounding regions, and thus vegetation would have become sparser and much more arid-adapted. One hypothesis is that Messinian-initiated North African aridity at this time could have been the origin of the Sahara, the earth's largest desert. Since we know that North Africa was well watered and densely forested in earlier (Oligocene and early Miocene) times, and that primates abounded there, the Sahara could well have played an important, but as yet unrecognized, part in evolutionarily dividing ancestral populations of African animals during the late Miocene.

In applying this scenario to the case of the hominids, the causal mechanism, climatic aridity caused by the Messinian Event, is in the wrong place for explaining their first appearance in the fossil record, which is eastern Africa. We must either bring the explanatory hypothesis to the fossils—that is, we must explain how Messinian-induced climatic aridity affected eastern Africa as well as North Africa—or we must bring fossils to the explanatory hypothesis—that is, we must find fossils of the earliest hominids in North Africa. Both efforts have been made.

BRINGING THE MOUNTAIN TO MOHAMMED: GLOBAL COOLING AND THE MESSINIAN EVENT

C. K. ("Bob") Brain is a scientific polymath, originally trained as a zoologist, but whose interests led him into detailed multidisciplinary investigations in paleoanthropology. His research on the South African cave sites, rich in early hominid fossils, in fact is a good example of how the traditional geological, paleontological, and anthropological disciplines have melded in paleoanthropology. It is not surprising from an intellectual standpoint that Brain published a paper in 1980 that related the Messinian Event to hominid origins. What was surprising was that this idea came from a part of the continent of Africa that was most distant from the locus of the Messinian Event. Brain's paper was a published version of an address to the *South* African Geological Society. How did he bring the metaphorical mountain of Messinian climatic change to Mohammed, the hominids, who from a standpoint of the fossil record resided only south of the Sahara?

Brain's approach was global. He related the Messinian Event to the pattern of global climate change, extension of ice sheets in Antarctica, and a worldwide decrease in sea level. Whether or not the Messinian Event in the Mediterranean was the initiator of climatic change, Brain thought that it was at least the harbinger of that change. Extension of the massive Antarctic ice sheet, however, made more sense as the mechanism by which climatic cooling would have occurred at the end of the Miocene. The water around Antarctica became colder and colder, and the cold Benguela Current that runs northward from the Antarctic Sea along Africa's west coast would have become progressively less evaporative. There would have been less rain over West and Central Africa, contributing to an overall decrease in forests and increase in savanna over large parts of the continent. As greater volumes of earth's water became locked up in ice on land, global sea level dropped. Atlantic water would have been prevented from flowing into the Mediterranean Basin over the Gibralter sill, which may have precipitated the gradual increase in salinity of the Mediterranean, leading to the Messinian Event.

In this way, Brain related the Messinian Event to general climatic shifts that in turn would have had a significant impact on hominid evolution. What became more important in Brain's model was the evidence for a worldwide oxygen isotopic shift to cooler temperatures at 5 million years ago. Paleontological evidence of course is needed to test this model, and 5-million-year-old fossil sites in Africa are not easy to come by.

BRINGING MOHAMMED TO THE MOUNTAIN: FOSSILS IN NORTH AFRICA

If Brain had approached the asymmetry of the Messinian hominid emergence equation from the standpoint of relating what had been thought to be regional climatic change to a global and pan-African pattern, I approached the question from another direction. Why not look at the evidence for Messinian climate change and hominid fossils at the same time, in the paleontological record, this time in the area surrounding the Mediterranean?

The best fossil site that I could find was located adjacent to a depression in the Libyan Desert shown on the map as Sebkha al Qayyanin. The site had been found in the 1920s by colonizing Italian soldiers stationed at a lonely desert fort originally built to guard the trans-Saharan caravan route. The fort was named Sahabi and it was by that name that the site was known. A famous fossil elephant skull had been found there and a restudy of the specimen had indicated that it was similar to, but somewhat more primitive than, a related species at Lothagam, Kenya. Unlike Lothagam, Sahabi was located within a few hundred kilometers of the Mediterranean and thus perfectly placed geographically to test the effects of regional climate change effected by the Messinian Event. It was the right age, with sediments beginning at perhaps 7 million years old and extending to younger ages. And the site was not only extensive but it had fossils that had been laid down by very slowly flowing water, raising hopes that relatively complete fossils could be found there.

Long-term fieldwork that I began with colleagues in 1978 has yielded a tremendous amount of data from Sahabi. The paleoenvironmental data, such as isotopes, sediments, fossil plants, and fossil animals, indicated that the climate in Messinian North Africa was indeed arid away from permanent sources of water. Sahabi presents a large concentration of animals and plants because it was deposited by a very large river. The question of hominids remains open, however. Fossils that we discovered at Sahabi documented the presence of two monkeys, which is generally accepted, and one hominoid, which is not yet generally accepted. We will return to the Sahabi hominoid later in this chapter. The importance of Sahabi at this stage of our research is that it has allowed the formulation of a hypothesis alternative to the Rift Valley/"East Side Story" Hypothesis, an expanded version of which was termed "Hypothesis 2" in the last chapter, to account for the chimpanzee–hominid evolutionary split.

THE SAHARA AS POPULATION SPLITTER

There are several ideas about how the Sahara came to be formed. In addition to the Messinian explanation, a popular hypothesis is that uplift of the Himalayas beginning in the middle Miocene resulted in a gigantic rain shadow that created both the Arabian and Sahara Deserts. Rain clouds caused by the prevailing winds from the east would have been blocked by the mountains. Some have pointed out that the uplift of the Himalayas seems somewhat too early for the first indications of desert in northern Africa. Both Messinian and Himalayan mechanisms may have been implicated in the formation of the Sahara, and much more data will be necessary to reach a conclusion. What is more important for our anthropogenic purposes is to understand the timing of the Sahara's formation and the nature of the associated climatic changes.

The modern plant biogeography of Africa reveals a very curious fact when it comes to desert plants. The two most important ancient foci of desert floras in Africa are southwestern Africa—the Kalahari and Namibian Deserts—and extreme northeastern Africa—the Ogaden Desert of Somalia. The Sahara, quite unconnected from either of these regions, does not retain any indication of an ancient focus for desert plant evolution, a sign that its extreme aridity is a geologically recent phenomenon. Prior to the discovery of the fossil plants from Sahabi, it was thought that the few European oak and laurel plants that grew on the tops of Saharan mountainous massifs and in the Canary Islands were remnants of the vegetation that once covered the Sahara. The abundant Messinian-aged fossil flora from Sahabi showed that this conception was entirely erroneous. North Africa had a vegetation of palms, acacias, and grasses essentially African and virtually identical to the extant sub-Saharan African flora.

Evidence from the deep-sea core off the west coast of Africa recently reported by Peter de Menocal of Lamont–Doherty Earth Observatory confirms that sand blown westward by the wind, indicating the presence of desert in the Sahara region, does not show up until the *late* Pliocene, that is, much later than the Messinian. We can be confident then in reconstructing the northern half of Africa in the late Miocene and early Pliocene as savanna and savanna woodland, but not desert. This is important because it demonstrates that the northern half of the continent was clearly a potentially habitable place for emergent hominids and has to be considered in hypotheses about hominid emergence.

The fact that the Sahara was a hospitable environment for early hominids during the Pliocene received dramatic support from the 1995 discovery by Michel Brunet, of the University of Poitiers in France, of a middle Pliocene, 3.0- to 3.5-million-year-old hominid fossil mandible in Chad. This specimen, nicknamed "Abel" after one of Brunet's colleagues, is located in what is now the virtual center of the eastern Sahara. Sidestepping for the moment the issue of where Abel's ancestors came from, it is clear that he and his population lived in what is now Chad in a habitat typified by large herbivores of tropical African affinities. It is a rather short deductive step to conclude that northern Africa would also have been habitable to even earlier hominids, perhaps even a population of our hypothetical hominid–chimp common ancestor. How might this creature have become isolated from its forest or forest-fringe environment and set off on its own evolutionary path in what is now the Sahara?

A glance at the map of modern vegetational zones of Africa (Figure 4) shows a marked horizontally striped pattern over most of the middle of the continent. There is a clear latitudinal gradient as one flies due north from the Congo forests. One passes through forest fringe woodlands, then into savanna with patches of woodlands and open grasslands, then into *sahel,* a mixed region of savanna grassland and arid-adapted scrub vegetation, and finally into the dunes of the Sahara. This east–west gradient of vegetation is immense, extending for more than 5,000 kilometers. It is a large front on which evolution may have advanced, and there is good evidence that the positions of the various belts of vegetation moved north and south as a result of the vicissitudes of climatic change through the Pliocene and Pleistocene.

The scale of these vegetational belts is so immense and homogeneous that one may well question how a forest population of hominoids might have been isolated on the other side (that is, the northern side). Specifically, how could a forest-fringe hominoid from one of the three forest refuges discussed earlier have ended up north of a southward-advancing front of savanna? There are two major mechanisms. The first is topographically related and the second is hydrographically related.

The Sahara underwent significant geological uplift at some time prior to the Pleistocene. Both the Tibesti and Ahoggar Massifs, in excess of 2,000 meters, were raised up in what is now southern Algeria and southern Libya. Because these areas of uplift mark the southern boundary of oil-containing beds, particularly in Libya, their age has been determined by oil geologists using a combination of marine paleontological correlations and absolute dating methods. Significant uplift of these Saharan mountains occurred in the late Miocene and Pliocene. The early-to-middle Miocene site of Gebel Zelten, in central Libya, thus dates prior to

Figure 4

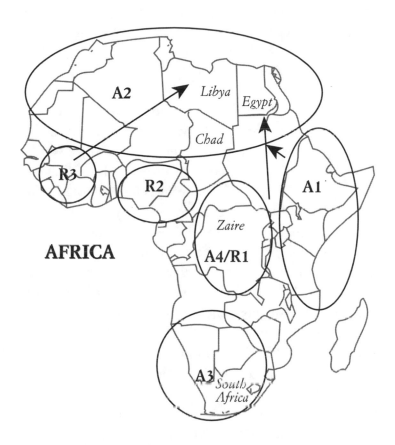

Africa holds records of both forest and savanna biomes. Forest refuges that today preserve ancient forest-adapted species of plants and animals are indicated by the regions R1, R2, and R3. These have both expanded and contracted as climates have changed in the past. Savanna habitats characterize the fossil sites known from regions A1, A2, and A3, but environmental fluctuations at different times have ranged from deserts to densely wooded habitats. Region A4 consists of fossil sites in the Western Rift Valley that in part sample ancient forests and some Plio-Pleistocene savanna habitats (from Boaz, 1996).

this uplift, and it is characterized by a terrestrial fauna with woodland and perhaps even forest affinities. As the Sahara became typified by more arid climate and more open vegetation in the late Miocene, denser vegetation would have retreated up the Saharan massifs, which may well have served as now-disappeared forest refuges. A good example of what some of these refuges may have looked like is provided by Mount Marsabit, now a national park in Kenya. This is an isolated volcanic mountain surrounded by the arid Chalbi Desert of northern Kenya, but clouds drifting in from the Indian Ocean serve to provide enough moisture condensing on the higher elevation to maintain a forest-adapted fauna here. Similar but much larger montane refuges likely existed in the Sahara throughout the late Miocene and Pliocene.

Routes of water courses were also of importance in animal and plant distributions. The Nile River now runs from tropical eastern Africa through the Sahara to empty into the Mediterranean. During the past the Nile would have sliced across all the latitudinal gradients, and its fringing gallery forests would have provided a conduit for populations to move back and forth between habitat gradients. There is also some evidence that a similar large river may have crossed from tropical Africa in the west to empty into the Mediterranean. The Niger, which still starts out as if it intends to cross the Sahara, makes a southerly U-turn to empty today into the Bight of Benin in western Africa. Before uplift of the Tibesti in Central Africa it may have followed a structural trough across the desert to empty into the Gulf of Sirt in Libya, accounting for the large source of tropical fresh water at Sahabi. Unlike the Western and Eastern Rift Valleys, which because of topographic relief can have no major water courses traversing them, the Sahara then had one, and possibly two, major migrational routes into its interior. Animal populations could move from the forested south into the Sahara savannas, but then become isolated by climatic and vegetational change—if an area equal to the size of the United States can be considered "isolated."

THE SAHARA AND HOMINID EMERGENCE

Unlike the African Rift Valley model of hominid emergence (our Hypothesis 2), the idea of a Saharan origin of the hominids is a new hypothesis. It is actually a series of hypotheses. Hypothesis 3 can be simply stated: The developing Sahara split hominids off from chimps.

What has been discovered that makes this theory of Saharan origins plausible? Why should we give it equal standing to the more venerable and generally accepted Rift Valley theory?

Geophysical dating of Saharan uplift and dune formation now indicates that the formation of the Sahara Desert as we know it was a much more recent phenomenon than many had guessed in the past. Its origins cannot date to before the Messinian Event. The first significant wind-blown detritus was not deposited until the middle Pliocene. Indeed, it is likely that most of the Sahara's extensive sand dates to the Pleistocene and Recent times. It should be remembered that the ancient Egyptians and Romans encountered a grassy and acacia-covered northern Sahara replete with hippos, giraffes, baboons, elephants, and zebras. Within only the last few thousand years, vast areas of Saharan savanna were turned into desert by human depredation, particularly poor agricultural practices and overgrazing.

Molecular clock estimates of the divergence of humans and chimps, discussed in the last chapter, accord well with the Messinian Event and initiation of the formation of the Sahara. The range of these estimates is 4 to 8 million years ago. Biomolecular analyses, however, show a consistent preference for ages at the younger end of that time span, that is, 4 to 6 million years ago. The formation of the African Rift Valleys, particularly the Eastern Rift, began and became significant physiographic features much earlier, in the middle Miocene. Our best estimate for the origin of the Western Rift is the late Miocene, approximately 8 million years ago, 3 million years earlier than the global cooling that caused aridification of the Sahara. Thus, a Saharan theory of hominid origins agrees with the biomolecular dates for hominid emergence better than does a Rift Valley theory.

The biogeographic beginning point for hypotheses on hominid emergence has to be the Central and West African forests. We must account for a spalling off of a part of this ancestral habitat, one that contained protohominids, if any model of hominid evolution is to make sense. The Rift Valley model (Hypothesis 2) does this by positing a north-to-south splitting off of part of the Central Forest Refuge that contained protohominids by the Western Rift Valley. The Saharan model (Hypothesis 3) suggests that this cutting off of a piece of tropical African forest containing hominid ancestors was caused by an east-to-west vegetational gradient from forest to savanna, with intra-Saharan forest or woodland refuges in regions of high topographic relief. In Hypothesis 2, hominids would have first evolved in eastern and connected parts of northeastern and southeastern Africa. In Hypothesis 3, hominids would have evolved in Africa north of the tropical forest belt and west of the African Rift Valley system. In Hypothesis 2, hominids would have emerged from the Central Forest Refuge. In Hypothesis 3, they could have emerged from any of the three forest refuges. The Nile provided a convenient conduit north from the Central Forest Refuge, and, prior to central Saharan uplift, tributaries of the Niger and the Lake Chad drainage may have provided other water and vegetational corridors through the Saharan savannas for a forest-fringe-adapted hominoid.

The Fossil Evidence for Hypothesis 3

As plausible as are the chronological, biomolecular, and biogeographical arguments for Hypothesis 3, the critical test for the model will come from the paleontological record. Here Hypothesis 3 is weaker than Hypothesis 2. After the remarkably diverse higher primate fauna from the Egyptian Fayum (discussed in Chapter 2), the North African fossil record of hominoid ancestry is virtually a blank through the entire Miocene and Pliocene. Only one Miocene site in the northern half of the continent, Wadi Moghara in Egypt, has yielded what appears to be an ape humerus. This find was only reported by Elwyn Simons in 1994. If this discovery is confirmed by further work, then it will support our paleoecological deduction that northern Africa was a largely forested place during the Miocene and a region hospitable to and inhabited by apes. I suspect that the dearth of ape fossils from the Miocene of northern Africa is a function of both past paleontological collecting strategies that focused on the large and "impressive" fossils and the extremely inhospitable conditions for fieldwork generally found in the Sahara today. Both circumstances can be remediated, even if they cannot be overcome entirely. The richly fossiliferous early-to-middle Miocene site of Gebel Zelten, central Libya, will be a particularly important site to test the predictions on northern African Miocene environments and hominoid evolution.

In the late Miocene–early Pliocene time period, there is the all-important fossil site of Sahabi, northern Libya, which straddles the Messinian Event. The three fragmentary fossils that I and my team collected from Sahabi that may represent hominoids have been the target of blistering criticism. I originally identified all three as unspecified Hominoidea, and I still believe after much more in-depth laboratory analysis that at least two of them are still best identified as such. My primary critics have been Tim White, an integrative biologist at the University of California, Berkeley, and his students. They believe that one of the specimens is from a dolphin, not a hominoid. They have not studied or commented on the other two specimens, a skull fragment and a fibula fragment. The heat that this controversy generated has served to keep most paleoanthropologists as far away from the Sahabi fossils as possible. But in 1988, Steven Ward and Andrew Hill ventured to study casts of the specimens for a review article on hominid origins that they were writing. In their opinion, the evidence was inconclusive to establish the presence of late Miocene–early Pliocene hominoids in northern Africa. There the matter has rested, and the original specimens remain in a drawer in my lab, largely ignored.

If White and colleagues are correct, and the Sahabi fossils do not represent a small hominoid in northern Africa 5 to 6 million years ago, then Lothagam and White's own site at Aramis, Ethiopia (4.0 to 4.2 million years old), records the earliest appearance of hominoids closely related to hominid ancestry in the fossil record. This result will tend to support Hypothesis 2, as advanced in the last chapter. If I am right and Sahabi does contain a record of a small-bodied hominoid in northern Africa at this time, this does not prove that it was a hominid but it does indicate that further research is needed. Sahabi, a site whose fossiliferous outcrops cover many square kilometers of the northern Libyan Desert, will be our most promising test of Hypothesis 3.

Complicating our test of Hypotheses 2 and 3, the abundant and well-preserved mammal fossils from Sahabi show that there were some clear connections with eastern Rift Valley Africa at the end of the Miocene. Pigs, antelopes, hippos, and elephants at Sahabi are closely related to Lothagam species but they are more primitive, indicating a somewhat earlier age for Sahabi than Lothagam. Unless there was some adaptive reason that a primordial hominid could not have ventured across the savanna intervening between Kenya and Libya, it stands to reason that hominids may be represented at both Sahabi and Lothagam. On the other hand, some elements of the Sahabi and Lothagam faunas are very distinct. Sahabi preserves abundant remains of large hippolike animals known as anthracotheres, relatives of which had died out in eastern African sites by the end of the early Miocene, 12 million years earlier. If the earliest hominids were tied to water or forest, as anthracotheres probably were, then there would be no reason to suspect that they would be represented at both sites. Only more research will tell, but even if hominoids are found to be present at both Lothagam and Sahabi, Hypothesis 3 holds that they should have emerged first in northern Africa. Can we deduce why this might have occurred?

New Views of the Messinian and Hypothesis 3

Hypothesis 3 gives us a scenario of hominid origins that involves displacement of a forest-living hominoid north of the African forest refuges. We have the sparsely documented presence of hominoids in the middle Miocene of Egypt, indicating that northern Africa was an appropriate ape habitat at this time, and we have the documented presence of hominids in the middle Pliocene of Chad, indicating that it was an appropriate hominid habitat at this time. The suspected hominoids from Sahabi represent a possible evolutionary link connecting the two datum points—the Miocene to the Pliocene. Assuming for the sake of investigating Hypothesis 3 that hominoids were present in the late Miocene and early Pliocene of North Africa, what evolutionary forces may have acted to transform them into hominids?

Recent geological research has shown that the early conceptions of the Messinian Event as a unitary and time-constrained event were overly simplistic. In 1993, Gian Battista Vai of the University of Rome and colleagues redated Messinian rocks in Sicily to 7.3 million years ago. This discovery is important because it shows that the Messinian was at least 1 million years longer, and that it started at least 1 million years earlier, than originally thought.

Other geologists who studied the deep-sea core rocks from the Mediterranean saw a recurring cycle in the sedimentary history of the deeper basins. There was not a single drying out of the Mediterranean, but up to fifteen episodes of desiccation. There were intervening periods of refilling of the basin, from either reflooding from the Atlantic or from increased inflow of fresh water from African and European rivers that empty into the Mediterranean. We now have a view of the Messinian as a period of earth history almost 2.5 million years in length, from 7.3 to 4.9 million years ago, that shows evidence of important climatic oscillations even though the overall trend toward cooler and drier conditions was dominant.

Thure Cerling of the University of Utah has shown from studies on the carbon and oxygen isotopes of carbonates in ancient soils that a major climatic shift occurred at about 7 million years ago. From sediments in Pakistan, Kenya, and even North America, he has found a significant increase in the amounts of a heavier form of carbon, C-4, used in the metabolism of many savanna plants, particularly grasses. Cerling and colleagues concluded that it was at this time that climates became drier and grasslands spread. In Indo-Pakistan the last apes went extinct. In Africa the effects of this climatic oscillation toward more arid habitats would have been felt first and most significantly at higher latitudes—in northern and southern Africa. The regional effects of Messinian desiccation and Himalayan uplift would have had an additive effect on the aridifying climate in northern Africa. All told, the recent evidence indicates that northern Africa was the area, adjacent to the forest habitats of African hominids, where oscillating climatic forces would have acted most forcefully in effecting evolution to an open-country adaptation in a protohominid population isolated in the Sahara savanna. Hypothesis 3 would thus be supported if the earliest hominids are discovered at sites dating to 6 to 7 million years ago in the Sahara. Eastern Africa should lack any evidence of hominids at this time and should only show the presence of hominids when they migrated in from northern Africa at the end of the Messinian.

ALTERNATIVE MODELS

On a subject as near and dear to the hearts of hominids as their own origins, there is bound to be a wide diversity of opinion. Most paleoanthropologists have given up on an Asian origin for the hominids, with *Ramapithecus* or *Sivapithecus* in the leading role. David Pilbeam of Harvard University, long a champion of this position, has abandoned it. Only Jeffrey Schwartz, of the University of Pittsburgh, continues to support this idea, but he does so against a mountain of contravening data from the biomolecular record and the anatomy of the fossils themselves.

Louis de Bonis of the University of Marseille in France has put forward a European origin for the hominids. He suggests that anatomical characteristics of the bony face and teeth of an ape known as *Ouranopithecus* from the Greek late Miocene show unique connections with hominids. These observations accord with other anatomical studies that tie a related European ape known as *Dryopithecus* to African hominoids, to the exclusion of Asian hominoids. These similarities may indeed indicate a closer evolutionary connection between the ancient Miocene apes of Africa and Europe than between either of these two groups and Asian apes. Any hypothesis, however, that suggests a unique origin of hominids from a non-African late Miocene source faces major problems when confronted with hominoid biogeography, paleoclimatic history, and the mounting hominid fossil evidence.

Finally, there is the "aquatic theory" of hominid origins, as resuscitated and popularized in recent years by Elaine Morgan of Great Britain. The idea was originally conceived by Sir Alistair Hardy, an Oxford astronomer, who was struck by a smattering of anatomical similarities between humans and marine mammals. Morgan expanded his argument and has attempted to tie the fossil record of hominids to a near-marine environment. Unfortunately for this theory, no evidence exists in the confirmed Rift Valley early hominid sites of any near-marine habitats. Even if marine habitats were nearby, as at Sahabi, there are many other much less fanciful interpretations of the adaptations typifying the earliest hominids. There is little in either the paleontological record or human morphophysiology to suggest anything other than a terrestrial existence since our fish ancestors left the sea some 350 million years ago.

HOMINIDS VAULT INTO THE PLIOCENE

The eminent African geologist and paleontologist Bill Bishop speculated in the late 1960s on the reason that the fossil record had not yet yielded hominid remains in the Pliocene. He concluded that it had been because the Pliocene had been an arid time, with little precipitation and relatively few fossiliferous sediments laid down by flowing water. We now know that hominid sites such as Laetoli, Hadar, the Middle Awash, and much of Omo and Turkana are Pliocene in age—that is, they date earlier than 1.8 million years ago. We also now know that they had a wetter and more wooded environmental context than later, Pleistocene sites, with the possible exception of Laetoli. However, viewed from the perspective of the Miocene, the Pliocene began with the aridifying effects of the Messinian Event and it indeed was more arid than conditions that had gone before. The relatively well-known portion of the hominid fossil record begins in the middle Pliocene, and it is to this span of time that we will turn in the next chapter.

5

Australopithecines Adapt to an Expanding Savanna-Grassland Environment

The year was 1924 and University of Witwatersrand anatomist Raymond Dart knew he had something unique between his hands. It was a fossil skull encased in rock that had been delivered to him the day of his daughter's wedding. He had been so excited that they had to drag him away to the ceremony. Now as he slowly and meticulously chipped away at the rock, dissecting out the delicate facial features of the skull, his mind reflected on how unique this specimen was. He had expected another baboon, but this was a hominoid—a member of that big-brained primate group that includes apes and humans. It had a small but globular braincase, a smooth browridge, a relatively small face, and a small canine, but he quickly reminded himself it was a juvenile and had not grown to its full size. Nevertheless, it was different from living juvenile chimps and gorillas. Dart used the specimen, dubbed the "Taung Baby," to name a genus and species new to science, *Australopithecus africanus,* in the February 11, 1925, issue of the British scientific journal *Nature.*

Dart's interpretation of the evolutionary significance of his *Australopithecus africanus* was quite conservative at first. He proposed that a new family, *Homo simiidae,* be erected for the inclusion of this single species, which he thought was intermediate between apes and humans. Only later and after further fossil discoveries did he promote the position that *Australopithecus* was the first hominid. His choice of a name in 1925 highlighted what Dart thought were the most unique aspects of the discovery, most resistant to later reinterpretation by other scientists: that it was some sort of ape ("pithecus" can actually refer to almost any primate from prosimian to hominoid), that it came from Africa, which was indisputable, and that it was from the south ("australo-")—also beyond dispute. Dart made much of the facts that the specimen was found far to the south of the known ranges of the chimpanzee and the gorilla in Central and West Africa, and that evidence of climate associated with *Australopithecus* proved the fossil animals were not forest species but those of the open-country South African *veldt,* a habitat into which the chimp and gorilla would never venture. In all the anthropological wrangling over the "hard evidence"—the anatomical characteristics and phylogenetic connections of *Australopithecus*—insufficient attention has been paid to these most fundamental aspects of the paleobiology of the australopithecines. This broader anthropogenic view is the subject of this chapter.

A RADIATION OF AUSTRALOPITHECINES

When Dart first discovered the Taung Baby and for a number of years afterward, the genus *Australopithecus* as we now understand it included only one real kind of australopithecine, one termed informally "gracile" because of its relatively lightly built skull and face. Its teeth were "harmonious" in their relative dimensions, with the front and back teeth about the same relative sizes as seen in later humans. Later a second australopithecine was discovered. This was the "robust" australopithecine, typified by a heavily crested skull, a face buttressed with such a massive amount of bone on the sides that the middle of the face seemed sunken in, and a "disharmonious" dentition, in which the huge chewing premolars and molars dwarfed the incisors and canines in the front of the mouth. Robust australopithecines are sometimes termed *Paranthropus,* but I use *Australopithecus robustus* to refer to the South African variety and *Australopithecus boisei* to refer to the East African species.

In recent years three more australopithecines have been added to the roster. *Australopithecus afarensis,* the species to which both the well-known "Lucy" skeleton and the trail of hominid footprints at Laetoli, Tanzania, belong, is similar to but more primitive than *Australopithecus africanus.* Its canine is relatively larger, its face projects more, its brain is smaller, and its skull sometimes has bony crests for the attachment of its chewing muscles. A hominid found in Kenya at around 4 million years ago was recently named *Australopithecus anamensis* by Meave Leakey and colleagues. It is the first hominid that we can say with certainty was bipedal. Finally, an even more primitive australopithecine, named *Ardipithecus ramidis,* has turned up in northern Ethiopia. Its teeth, which are the only body parts well known enough to make any deductions about this intriguing creature, are very similar in size and even anatomy to the living bonobo, although they are clearly hominid in their flattened crowns and nonslicing canines.

Figure 5 illustrates my assessment of the pattern of shared anatomical similarities (a *cladogram*) and the evolutionary tree (a *phylogeny*) that I derive from this. Other specialists may divide up the fossils and classify them somewhat differently. For the reader with a more specialized interest, references with alternative interpretations are included in the bibliography. Although the details are of abiding fascination to the anatomical paleoanthropologists who study them, they are of less concern to us here. We seek to understand the ecological relationships of the major groups of australopithecines and how they evolved.

The australopithecines represent a unique phenomenon in latter hominoid evolution—an expansion of species diversity and range. These first generally recognized hominids expanded because the environments that they inhabited expanded. It is the central thesis of this chapter that the gradual opening of the vegetation over much of Africa offered evolutionary opportunities for hominids that contributed to their evolutionary expansion. All other hominoids of which we are aware at this time either died out or managed to stave off extinction by remaining in isolated forest refuges.

A number of hypotheses on the anatomy, physiology, ecological adaptations, paleoclimatic history, and fossil record of the australopithecines are subsumed into Hypothesis 4 in this chapter, which can be summarized as: Australopithecines adapt to an expanding savanna-grassland environment. This theory, unlike some of the others presented in this book, is generally well accepted. But there have been both past criticisms of the "savanna myth," as well as some new discoveries, that have thrown doubt on it. We shall first set the paleoclimatic stage, then look at hominid comparative anatomy and physiology, and finally discuss in some detail the fossil evidence for australopithecine adaptations. We shall see that the evidence supports the characterization of australopithecines as the only "savanna apes."

Pliocene Paleoclimate and the Spread of the African Savannas

In the Pliocene, Africa saw habitats with fewer trees and extensive grassland spread widely over the northern, eastern, and southern

Figure 5

CLADOGRAM

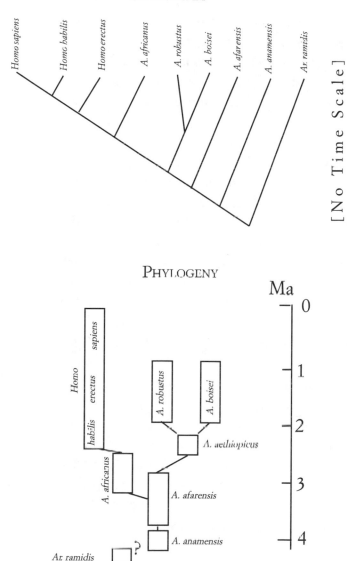

The phylogenetic framework used in this book. The figure at the top is a cladogram, which simply places a number of species in an ordered array of similarity, based on uniquely derived anatomical "character states." The figure at the bottom is a phylogeny derived from this cladogram, which shows hypothesized evolutionary relationships of species and their relationships through time. (Ma = millions of years ago.)

parts of the continent. This habitat type became more common than at any preceding time in the Cenozoic era. Ecologists have debated for years over the correct terms to use to denote these types of environments. *Savanna,* originally a Caribbean Indian word, has endured as a descriptor of these typically African ecological zones despite the imprecision of the term and its geographical origins in the Western Hemisphere. Perhaps its very imprecision accounts for the term's continued use because paradoxically it accurately describes the variability of these habitats in nature. Savannas can be open grassland, when they can be qualified as *treeless savanna, grassland savanna,* or *steppe.* They can have a variable amount of tree cover, and the terms *wooded savanna* and *savanna woodland* have evolved to refer to these habitats. In between these two extremes of grassland and savanna woodland, there is savanna that does not have trees but has a lot of shrubs and bushes within an overall matrix of grassland. This third overall category is called *shrub savanna* or *bush.* All savanna, however, has grassland as a component of its ecological profile.

We know that savanna environments spread in Africa during the Pliocene because of isotopic changes in both the deep sea and land records, and from direct paleontological documentation of changes in flora and fauna at African fossil sites. The oxygen isotope curve derived from data from the Deep Sea Drilling Project now gives us a clear view of global temperatures during the Pliocene (Figure 6). Starting at a high at the beginning of the epoch, there is a gradual decrease in average global temperatures as we approach the Pleistocene, at around 1.8 million years ago. From a number of different sources we can deduce that this temperature decrease corresponded to an average decrease in rainfall over much of Africa. African vegetation responded by retreat of forests, which are dependent on high levels of rainfall, into zones of low barometric pressure such as the Congo Basin or up mountains and areas of high relief.

The isotopic record of the Pliocene also shows, in addition to

Figure 6

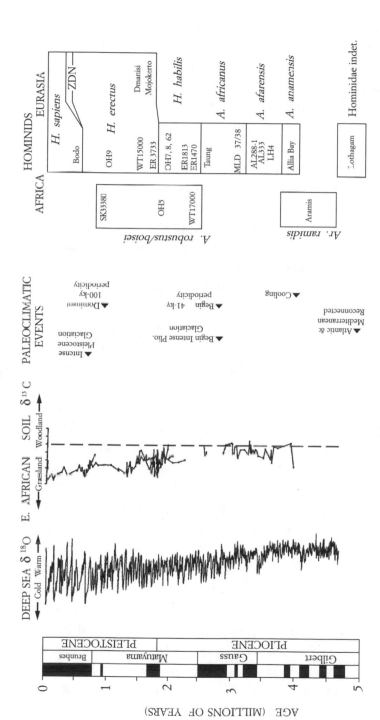

Plio-Pleistocene paleoclimate and hominid evolution. The oxygen isotope curve shows a mean trend toward colder temperatures as one approaches the present, with a second trend of increasing amplitude (greater fluctuations) through time. Much discussion has centered on paleoclimatic "events" at 5 ma, ca. 2.5 ma, and ca. 1 ma and their effects on hominid evolution. (After Boaz and Burckle, 1984, and de Menocal, 1995). (Ma = millions of years ago.)

the average changes in global temperature, that temperatures began to fluctuate more widely beginning in the Pliocene. The graph of change shows progressively more scatter around the mean when we enter the Pliocene from the preceding Miocene, and as we go from the early to the late Pliocene. This scatter around the mean is not "noise" but an indication that the cold periods got colder and the dry periods became drier. Despite the overall trend, the intervening hot periods got hotter and the wet periods were rainier. Extrapolating to shorter time scales for which we do not have detailed data, the differences within one year from one season to the next would have become more marked. Earlier times had been more stable. This aspect of climatic and vegetational change in the Pliocene was probably much more important from an evolutionary standpoint than the overall "average" decrease in temperature and rainfall, which we can only see from our vantage point several million years later.

The fluctuations in climate and what causes them have been a topic of intense scientific interest for many years. A Serbian astronomer named Milutin Milankovitch proposed in the 1930s that cycles of the solar system and the relative amounts of sunlight hitting the earth account for much of the driving force for climatic fluctuations. Because the orbit of the earth around the sun is not a perfect circle, but instead an oval, the earth will get more incident sunlight when it is in the short end of the oval, near the sun. This point of maximum incident solar radiation is called *perihelion* (for "near the sun"). The movement of the earth around its orbit is called *precession*. Another important factor is the wobble of the earth as it rotates on its own axis. At different seasons of the year the tilt of the earth's axis accounts for different amounts of sunlight hitting the Northern and Southern Hemispheres. Since precession and the seasonal tilt of the earth's axis occur on different and independent time scales, calculations must be undertaken to ascertain their combined effect at one point in time. If Milankovitch is right, the resulting

fluctuations in solar radiant energy hitting the earth should show
a regularity and predictability in climatic patterns through time.
With a number of caveats, they do.

In Africa during the first three-quarters of the Pliocene there is
a cycle of dry and wet periods that recurs every 19,000 to 23,000
years, a length of time corresponding to a Milankovitch cycle.
This pattern of climate periodicity is picked up in the deep-sea
core records of wind-blown dust off the coasts surrounding
Africa. It is a pattern that stayed stable in amplitude for some 14
million years, from the middle Miocene until the late Pliocene.
The global precessional pattern translates into a pattern of rainfall
over most of Africa during the summer monsoons, when winds
from the Arabian Sea and Indian Ocean bring water-laden clouds
inland.

The oxygen isotope record shows that despite the continuity in
pattern of rainfall over this vast span of time in Africa, the *amount*
of rainfall decreased. Climate was becoming drier from the
Miocene into the Pliocene. The australopithecines were hominids
that became well adapted to this predictable seasonality of rainy
and dry seasons.

In general terms, Africa during most of the Pliocene was a
place and a time of gradually increasing dryness of habitats with
expanding dry-adapted vegetation. The paleontological record
locates hominid emergence within this time frame and place.
Thus, it is reasonable to look for, in the first instance, a connection between the ecological changes and the evolutionary appearance of hominids. Most lineages of mammals that we find in
the Pliocene fossil record of Africa show progressive adaptations
to these more open conditions. Antelopes and pigs, for example,
show increases in height and length of their molar teeth, adaptations for eating tough savanna plants such as grasses. We must
make sense of the adaptations of the australopithecines from a
standpoint of life in the savanna if Hypothesis 4 is to hold up to
closer scrutiny.

ECOLOGICAL LIMITS
AND THE RISE OF THE SAVANNA APE

In looking at what ecological factors explain the distribution of animals, biologists have found that the "rule of maxima and minima" is very important. Paleontologists have made the same discovery in the fossil record (which we shall discuss in the next chapter). This rule states that the important factor that limits an animal's possible range is the one environmental variable that it cannot tolerate. A tropical mouse, for example, may be limited to near the equator despite the presence of appropriate habitats farther north and south, because in some part of its life cycle it cannot eat, nest, breed, or survive when temperatures at night get below a critical value. This critical minimum value may only be reached one or two nights a year, and in some years not at all, but that is enough to keep the population of mice from being able to live long-term at higher latitudes. In other species the critical variable may be availability of daily drinking water, rocky places to hide, abundance of a particular food plant, or virtually any other aspect of the environment that is tied into the species' adaptation. In reality, there is a complex web of ecological interactions between a species' adaptation and the environment. However, by focusing on the single minimum or maximum variable that defines a species' range, a scientist is able to hone in on the unique or most characteristic component of a species' adaptation—the one that separates it from other closely related species.

In examining the ranges of the African hominoids, there is one overarching ecological dichotomy between their habitats. Of the four species still alive in Africa today, three (*Gorilla gorilla, Pan troglodytes,* and *Pan paniscus*) are forest-living, and only one, *Homo sapiens,* is savanna-living. It is granted that the forest habitats of the latter two species range into more "open" forest or forest fringe, the point made in the previous chapter, but it is forest nonetheless. And humans do, informally speaking, "live in forests." This apparent exception, however, offered by the large populations in mostly forested countries like Zaire, is only an exception at a distance. On closer inspection, most people can only live in dense forest when they clear little open patches of savanna, on which they build a house and plant a garden. Furthermore, recent studies on "pygmies," such as the Mbuti of Zaire, who do live in dense forest with minimum modification of the environment, indicate that they cannot subsist on food from the forest alone. They must rely on barter with agriculturalists in order to maintain an adequate diet. In spite of some blurring at the edges, the distinction between forested ape environments and nonforested human environments is a real one.

The specific ecological minima or maxima that keep Tanzanian chimpanzees from roaming from their forested enclave at Gombe over to the savanna of the Serengeti Plain, only about 250 kilometers distant, correlates quite closely with the density of trees. To the west of Gombe, where trees extend in a continuous blanket of forest, the distribution of *Pan troglodytes* ranges over 5,000 kilometers, to Liberia. Ecologically speaking, what is it that specifically keeps chimps out of the savanna? We do not know, but whatever it is, there is a good chance that it is closely related to what changed in hominid evolution to allow our ancestors, savanna apes, to take advantage of an expanding habitat type in the African Pliocene.

Diet is the front runner in choices for a limiting factor in chimpanzee range. Chimps eat a lot of fruit, and if fruiting trees are too few or too dispersed on the savanna, they cannot gather enough in a day's time to stave off hunger. Chimps also eat meat, and there is certainly much more meat on the hoof and paw on the savanna than in the forest. Evidence indicates, however, that chimps do not have the teeth or gastrointestinal tracts to adequately digest meat. Although they bolt their meat with relish when they can get it, they do not chew it well. It frequently passes through their systems without being adequately broken down and absorbed. Chimps thus engage in the disgusting habit (from a human perspective) of picking through their feces and re-eating portions of undigested meat, a practice termed *coprophagy*. If chimp digestive systems cannot handle one of the most prolific food sources in the savanna, perhaps this is a limiting factor in their dispersal. Chimps also ingest a lot of water, both in their succulent fruit diets and through direct drinking of water, sometimes with crunched-up leaf "tools." Water sources on the savanna are few and far between, and perhaps it is thirst, not hunger, that limits chimps to their cool forests.

A second ecological parameter that may be limiting to chimps making a run onto the savanna is heat load. As everyone knows, chimps have black hair all over their bodies. A black surface absorbs heat from incident sunlight, rather than reflecting it. Thus, the amount of heat absorbed by a black-haired chimp on a cloudless day in the savanna is much greater than the amount of heat absorbed by the same chimp in the shade of the forest, even if the ambient air temperature is the same. The chimp's heavy coat of hair also insulates its body from losing heat by convection with the surrounding air currents. A chimp therefore must avoid getting overheated or run the risk of dying from heat prostration. In the forest the chimp's hair functions to keep it warm when it is cold, especially at night, relatively dry when it is raining, and to protect its skin from insects and from minor injury. The hotness and dryness of the savanna, the type of habitat in which people tend to do well (especially if they have a history of sinus infections), may then actually be the effective barrier to chimp dispersal out of the forest.

A major difference in the habitats of forests and savanna is the substrate, the ground surface. Adult chimps spend a good part of time on the ground, and they can be considered terrestrial quadrupedal animals, although their adaptation still clearly includes climbing and living in trees. The locomotion of chimps is "geared" much lower than that of humans. Chimps move powerfully through uneven terrain, much the same as a four-wheel-drive Landrover or Hummer, and they use a lot of energy in doing so (my Landrovers never got over 12 miles to the gallon).

Humans are much more like small passenger cars built for flat highway driving—they use relatively less energy to traverse long distances—but they are useless for rough terrain and inefficient when powerful locomotion is needed. Confronted with the long vistas of the savanna, chimps may anticipate that they will run out of gas long before they reach the far horizon, and they turn back to the chimp-scale, uneven habitats to which they are familiar. Recalling John Van Couvering's cartoon, it is even possible that chimps may find that the hot ground of the savanna burns their knuckles! For whatever reason, they do not venture far from wooded havens when they are at the fringe of their forest habitats.

Humans, in contrast to chimps, have been called "linear in build, large, meaty, hairless, and sweaty" (by biological anthropologist R. W. Newman). They also bite hard and walk upright. Human physiology is quite different in a number of respects from that of chimps, and the anatomy or presumed anatomy of the earliest hominids has been interpreted in light of these physiological differences. We will now turn to the theories that make up Hypothesis 4, which offers a scenario of australopithecines adapting to the open habitats of the African savanna.

"Ye Shall Live by the Sweat of Thy Brow"

When hominids left the verdant Garden of Eden of the African forest refuges, they began to sweat. Sweating is one of the biggest differences that separates human physiology from that of chimps. Yet it is one of the most paradoxical. If hominids needed to conserve their bodies' supply of water in a hot and dry habitat, why evolve an adaptation that secretes up to a liter of water an hour through the pores of sweat glands in the skin? The answer that generations of physiologists and anthropologists have offered is dissipation of heat load. Hominids lost their hairy coats and became "naked apes" at the same time that they became savanna apes.

Their skin became an organ of heat regulation. When internal body temperature or skin temperature becomes elevated, due to either physical exertion or the sun's heat, watery sweat pours out onto the skin. Circulating air currents evaporate the sweat from the skin, a process that cools the skin's surface and pulls heat out of the body.

Earlier theorists suggested that sweating was an adaptation for hunting in the middle of the day, a time when hominids, relatively slow of foot but with their superior cooling mechanism, could effectively run down prey. Because almost all other animals lack similar sweating mechanisms, they would succumb to the heat after a short chase. The middle of the day was also a time when hominids could avoid competition with other savanna predators, such as lions, leopards, hyenas, cheetahs, or wild dogs, which hunt in the cool crepuscular hours of dawn and dusk. As persuasive as this hypothesis has been, it may also be an overstatement. Hominids did not live by meat alone, so running down game was not the only, or even primary, means of obtaining food.

Modern Africans who live in the savanna can do quite well without physical exertion in the middle of the day. It is likely, more often than not, that australopithecines on the savanna would have found a nice shady acacia to sit under at noontime. I can personally attest to the wisdom of this approach. Much more work can be accomplished, be it gathering or hunting (or in my case, surveying for fossils), during a three-hour span in the cool morning or evening compared to a similar period of time in the middle of the day. Australopithecines probably *could* have run down a small antelope in the middle of the day, but only if they had to. Perhaps this is the primary importance of this physiological adaptation. It changed the ecological minima and maxima under which hominids could subsist. Sweating certainly lowered the ecological barrier that the higher temperatures and greater incident sunlight on the savanna had posed for forest-living hominoids, allowing some of their descendants to colonize this expanding habitat during the Pliocene.

If we can now reconstruct australopithecines as hairless, sweaty savanna apes living in a hot environment, there is still the fact that the savanna is also a *dry* environment. Where did the australopithecines replenish the water that evaporated from their skin as sweat? Even today, our kidneys reveal the fact that we are only recently removed from an adaptation to a well-watered forest habitat. We excrete a lot of water in our quite dilute urine. Earlier colonizers of the savanna, such as dogs, rats, and especially birds, are all able to wring out much more water in their kidneys and produce very concentrated urine compared to ours. We can thus conclude that australopithecines needed to drink water every day and had to live in those parts of the savanna within a day's walk of water. Like modern Africans, they may have dug in ephemeral sandy streambeds for water, eaten water-laden succulent plants growing in the savanna, and even carried water in some sort of container. If chimps can crumple a leaf and use it to drink up water from the bowl of a tree, it is not too difficult to postulate that australopithecines could have fashioned a leaf, gourd, egg shell, or similar container for water. A daily need for water thus tethers hominids to savanna habitats with predictable supplies of water.

ALFRESCO DINING IN THE SAVANNA

What new types of food would forest-living hominoids moving to a life in the savanna need to be able to eat? We have already mentioned meat. Chimps like meat, but they also do not do a good job of digesting it. Reingesting food that has already passed through the digestive system once is obviously not an efficient food extraction mechanism. Australopithecines could have adapted to making meat more digestible in one of two ways: by evolving strong stomach acidity to actively break down meat that was swallowed in large chunks or unchewed pieces, a strategy used by carnivorous vertebrates from crocodiles to lions; or by comminuting the meat into tiny bits in the mouth so that it would be ready without much more modification for absorption in the small intestines. The australopithecines evolved to utilize the latter approach.

Australopithecine skull and dental anatomy show that these hominids had big, flat, chewing teeth—premolars and molars— that in comparison to their body size were huge, unlike the cheek teeth of chimps or gorillas. The tops of these teeth had thick layers of very hard enamel that could withstand both a lot of abrasion from tough and fibrous food, which would have worn them down, and heavy chewing forces, which would have tended to crack their surfaces and break them apart. The muscles of mastication in the australopithecines moved to the front of the skull, so that these muscles were at a mechanical advantage in generating the maximum force for their contraction right over the cheek teeth. Hominids can bite down harder with their molars than gorillas, who with their massive chewing muscles located mostly toward the back of their heads exert the greatest bite force on their front teeth.

Australopithecine teeth then allowed them to chew up and digest uncooked meat, which is extremely tough and fibrous, thereby opening up a major food resource on the savanna unavailable to small-molared forest-adapted hominoids. But the remainder of the australopithecine ancestral diet, perhaps the bulk of it—plant foods—would also have been significantly impacted by an ecological move to the savanna. Whereas forest plants and fruits tend to be succulent and thin-skinned because they live in a humid environment with abundant water, savanna plants tend to have thick skins to prevent water loss through transpiration and to discourage insects and other animals from attempting to utilize them as water sources in a dry environment. Thick-enameled, large chewing teeth would have been an important component of australopithecines' ability to eat and digest even the plants that grew on the savanna.

THE ECOLOGY OF BIPEDALISM

Since antiquity, when Aristotle and his fellow philosophers decided that humans were best defined as "featherless bipeds," an upright stance and all the anatomy that goes along with it have occupied a central place in discussions of human origins. The modern scientific debate on bipedalism goes back, as does so much else in evolutionary biology, to Charles Darwin. Darwin postulated that bipedalism evolved in the human lineage because it freed the hands from their function in locomotion and thus enabled tool use, the primary means with which modern humans adapt to the environment (or adapt the environment to themselves). Darwin's idea has fallen into disfavor in recent years mainly because the fossil record now shows that hominid bipedalism appeared by about 4 million years ago and there is no evidence of hominid tools until 2.5 million years ago. I have pointed out in other writings that this "test" of Darwin's hypothesis on bipedalism is not definitive because it is based on only the *stone* tool record. Australopithecines, like chimps, probably made many kinds of nonstone tools before the use of stone was discovered. And some of these types of tools, apparently made from bone by hominids and not modified by other animals or geological processes, have recently turned up in the cave of Swartkrans in South Africa. I suspect that with further paleoanthropological digging, and with a more extensive knowledge of nonhuman primate tool use, we will discover that the australopithecines were accomplished tool-users. I do not agree with Darwin, however, that tool use was the evolutionary cause, sometimes called the "prime mover," for hominid bipedalism.

Hominoids ancestral to the australopithecines living in the forest fringes of Central Africa already possessed the ability to fashion and use tools, mostly for food-getting purposes, if we take chimp tool-making abilities as a baseline. If their environment had stayed the same, those hominoids probably would still be there, and in a sense they are, in the form of the common chimp and bonobo. Undoubtedly, chimp manual object manipulation, carrying, and tool use would all be significantly enhanced if the species gave up knuckle-walking and evolved bipedalism. But knuckle-walking seems to have served the chimp admirably well in its environment, and there is no indication that natural selection is pushing our ape cousins toward an adaptation more like ours. Specifically, the fact that chimps make and use tools does not seem to be a strong correlate of bipedalism.

Ecological change to widespread savanna environments and adaptation to those habitats were, I believe, the more basic changes that set into motion the forces of natural selection that led to hominid upright walking. This idea of an association of an open "plains" habitat with human evolutionary emergence and bipedalism is also an old hypothesis. It was first proposed by the codiscoverer of the theory of evolution by natural selection, Alfred Russel Wallace, who suggested that hominids had first arisen on the Central Asian Plain. Wallace was mistaken about the geography of human origins but he may very well have been right about the ecological correlate to bipedalism.

Human anatomy of the hip, leg, and foot demonstrates that we are adapted to walking, striding, and running on substrates that are largely flat and even. This was a major part of Wallace's argument. Our feet have lost the grasping ability and the opposability of the great toe of ape feet. In Wallace's day the apes were classified in the *Quadrumana,* a primate category that separated the apes from hominids because they have "four hands"—that is, hands *and* feet that could grasp an uneven forested or arboreal substrate. Initially, there were no early hominid fossils to test Wallace's hypothesis. Now there are quite a few australopithecine skeletal parts that tell us about the adaptations of their limbs, and it is clear that they were bipeds. Some of the newer fossil evidence requires, as we shall see, some modifications to the straightforward correlation of savanna environments and hominid bipedalism, but in large part Wallace's idea remains intact.

In looking at bipedalism within a context of comparative

anatomy and physiology, biologists have noted for some time that
it is a very unusual way for a vertebrate animal to get around. Virtually every other back-boned animal moves through its habitat,
be it sea, land, water, or some combination of the three, with its
body parallel to the surface of the earth. Some of the major excep-
tions are the bipedal dinosaurs (their descendants, the birds, also
tend to walk and hop on two legs when they are on the ground),
kangaroos, and a tiny fossil insectivorous mammal recently found
at the German Eocene site of Messel. Bipedalism in dinosaurs
seems to have been most likely a predatory adaptation, raising the
head up and allowing better sighting of prey. Later birds became
obligate bipeds because their forelimbs evolved their unique specializations for flight. Kangaroos, on the other hand, being
pouched mammals or marsupials, may have evolved bipedal hopping as a way to move rapidly but at the same time keep their immature and helpless offspring from falling out of the protective
pouch. Why an extinct tiny insectivore would have evolved
bipedalism has so far eluded theorists' imaginations. None of
these types of bipedal locomotion seem particularly analogous to
the hominid condition. But because bipedalism is a rare form of
locomotion, its evolutionary appearance in the hominids has required explanation. Primates have been the animals that have
served scientists as comparative models from which to theoretically derive hominid bipedalism.

The British anatomist Frederic Wood Jones, at the turn of the
century, postulated on the basis of his comparative anatomical
studies that the small prosimian primate, the tarsier, was the primate most closely similar to the ancient forerunners of humankind. A major part of Wood Jones's argument was that
tarsiers' anatomical structure had become modified to an upright
position because of the way tarsiers sat perched on the sides of
trees before their tremendous leaps to catch insect prey. No fossil
evidence ever surfaced to support Wood Jones's ideas, but his
anatomical observations did make sense if applied to primates
more closely related to humans, the apes. Other anatomists

pointed out that apes, unlike quadrupedal monkeys, had trunks already adapted to supporting the upper body vertically because apes had evolved to hang and climb in trees using their widely extendable upper limbs. The only significant anatomical changes necessary to convert a tree-living ape into a ground-living biped were lengthening of the legs and development of knee, ankle, and foot support mechanisms.

Studies on modern chimps were carried out to determine under which conditions these apes showed the most bipedal behavior. All the situations have potential relevance to the selective reasons that hominids may have evolved bipedalism as a primary means of locomotion. Chimps walk bipedally when they are carrying objects—particularly food objects. They adopt bipedal stances when they are aggressively confronting each other or a potential predator, when they frequently brandish branches and throw objects. The argument has been made that early hominids would have been at a permanent advantage and would have appeared "larger" to potential predators on the savanna if they were habitually bipedal. Chimps, as well as their savanna-living cousins, the baboons, stand upright to look over tall grass, frequently climbing small elevations to do so. A bipedal stance thus confers a much better view of the surrounding environment, important both for foraging and for avoiding predators. Prairie dogs in the American West can be observed in the same type of behavior. Chimps have even been observed walking bipedally to avoid getting their hands wet in the morning dew.

All these observations are useful in deducing aspects of hominid bipedal adaptation to the savanna, and the behavioral aspects of that adaptation. But I would maintain that they are all secondary to the ecological prime mover of increasing aridity and geographic isolation that brought about selection for bipedalism in the first place. There would be no need for carrying objects, warding off predators, having a better field of vision, and even keeping one's knuckles dry in the grass if the initial move to the savanna had not already taken place.

Physiologist P. E. Wheeler has made a series of observations relating to the ecological advantages of bipedalism in minimizing heat load in a savanna biped. In addition to the other adaptations to dissipating heat—sweating and hairlessness—Wheeler points out that the "footprint" of shadow or total body area presented to the sun is much less in an upright biped than in a quadrupedal, knuckle-walking chimp. Only the head and shoulders receive the brunt of solar radiation and thus the total amount of body surface absorbing radiant heat is reduced. Anthropologist Dean Falk has extended this idea to suggest that early hominids may even have had a mechanism of cooling the head and brain by shunting blood though emissary veins in the scalp and thus dissipating heat. Some have termed this the "Radiator Brain Hypothesis." Whether this physiological mechanism turns out to be a major heat dissipator, it is true that all modern humans have their greatest amount of body hair on the tops of their heads, and the most likely adaptive explanation for this placement is that it protects the head from solar radiation. Reducing the footprint of heat absorption from the savanna sun is not likely to have been the prime mover in effecting the move of hominids to the savanna, but it certainly was an important consideration once they got there.

Savanna and the Fossil Record of Hominids

In broad climatic terms, the hominids arose in Africa during the gradual onset of drier habitats—savannas. But the detailed records of paleoclimate from the fossil sites that have yielded the hominids can further elaborate and refine our understanding of how the hominids became adapted to the savanna.

The earliest records of australopithecines for which we feel confident enough to place species names are from Aramis, Ethiopia, and Allia Bay, Kenya. Both are about 4 million years old. The former site has given us some teeth and fragmentary bones that have been named *Ardipithecus ramidis* (meaning "stem earth-ape") and from the latter site we have *Australopithe-cus anamensis* (meaning "lake"). Interestingly, both sites preserve fauna and flora indicative of forest, not savanna, habitats. Aramis has a large number of forest rodents and colobine monkeys and almost no savanna-adapted antelopes. Allia Bay preserves forest plants that today are found only in the forests of Central and West Africa. These new data will be seized upon by some to argue against the correlation of savanna and early hominids. I disagree, although I think the data from Aramis and Allia Bay add signifi-cantly to our understanding of how early hominids initially adapted to the savanna.

There are several other sites in Africa from which the earliest hominids have been discovered and which either show a mixed savanna–forest habitat or an indication of much more open condi-tions than either Aramis or Allia Bay. Lothagam, which I accept as the earliest record of Hominidae, is about a million years earlier than Aramis or Allia Bay. I reviewed the fauna and available paleo-climatic indicators from this site a number of years ago and con-cluded that they represent a mixed savanna environment with some trees, but definitely not a forest. There are a number of ante-lope species present and the overall predominance of large-bodied mammals at the site is a pattern that we would expect on the sa-vanna.

Younger than Aramis or Allia Bay is the site of Laetoli, Tanza-nia, which has yielded remains of *Australopithecus afarensis*. This site not only preserves a clear, open savanna fauna, but records the passage of three hominids over a largely denuded landscape cov-ered by volcanic ash. The Laetoli trails of footprints have attracted much attention because they unequivocally show hominid bipedalism, but they demonstrate equally well that hominids were eminently capable of traversing open savanna. Laetoli dates to 3.6 to 3.8 million years ago, only about a quarter of a million years older than Aramis and Allia Bay.

I conclude from this evidence that the earliest hominids were capable of walking across significant expanses of savanna, an ecological "minimum" that chimps cannot overcome, but that they were not yet adapted to life on the open plains of Africa. They were most at home in their ancestral habitat of trees, from which they obtained sustenance in the form of fruits, shade from the sun, protection from predators by climbing their branches, and most probably their first tools. These were habitats that chimps would have found habitable, if unreachable.

Later sites that indicate the presence of australopithecines, such as Omo, Hadar (near Aramis), East and West Lake Turkana (near Allia Bay), Olduvai (near Laetoli), and the South African cave sites, all show that savanna habitats increased in representation during the middle and late Pliocene. Hadar was the most heavily vegetated of these sites, but the mammal fauna even here, unlike Aramis, was a large-bodied savanna fauna. There would have been few or no actual forest patches within which australopithecines could have sheltered. During the first half of the Pliocene, natural selection would have progressively pushed the hominids toward a more exclusively savanna-oriented existence.

Anatomy supports the interpretation that the earliest hominids were still closely tied into trees, even if those trees were surrounded by large expanses of savanna. All the australopithecines had powerful forelimbs with long fingers that possessed heavy flexor muscles. They also had long toes, including a great toe that was somewhat divergent from the rest and could have possibly been used in grasping onto tree trunks. The most reasonable interpretation of this anatomical suite of characters is that it was part of an adaptation for habitual climbing, but one that coexisted with the australopithecines' ability to walk bipedally.

Two Kinds of Australopithecines

Since the early days of paleoanthropological discovery in South Africa, it has been apparent that there were two main "kinds" of australopithecines. One of these, the first to be discovered, was a smooth-skulled little hominid with teeth more or less like ours. The Taung Baby and other *Australopithecus africanus* from South Africa were termed "gracile" australopithecines. A second kind of australopithecine was discovered at the South African sites of Kromdraai and Swartkrans. This species had a prominent crest of bone running along the middle of its skull, and massive chewing teeth. Its face was sunken in front because of the relative reduction of its front teeth and the large amount of bone that developed on the sides to support its premolar and molar teeth. These hominids were called "robust" australopithecines.

In eastern Africa there seem to be the same two kinds of australopithecines—gracile and robust. Similar hominids, particularly "Lucy" and other *Australopithecus afarensis,* probably represent the gracile australopithecines in eastern Africa. For the most part they are somewhat more primitive and earlier in time than the South African *Australopithecus africanus,* a species that is poorly represented in eastern Africa (probably due to chance and a lack of localities exactly the same age as those in South Africa). There is, however, a good record of robust australopithecines in eastern Africa, and even though they are usually referred to as a species different from *Australopithecus robustus*—the South African species—there is consensus that the two are closely related.

How, when, and where the gracile and robust australopithecines differentiated has never been determined. The hypothesis that I offer here I first proposed in 1977. Fossil discoveries since then have helped to refine the dates and to support the model.

The robust australopithecines have an anatomy that is uniquely derived for the hominoids, and they therefore must have evolved from a more generalized gracile species, one that looked like *Australopithecus afarensis* or *Australopithecus africanus.* The earliest robust australopithecine is a skull from West Lake Turkana, Kenya. It is known as the "Black Skull" because of its heavily pigmented surface. It dates to 2.5 million years ago. The ancestry of the Black Skull is unknown because no fossils exist that clearly show its evolution from either *A. africanus* or *A. afarensis.* Most paleoanthropologists believe, however, that the robust australo-pithecines evolved somehow in eastern or southern Africa from gracile australopithecines of some description whose fossils have just not been found yet.

I think that this scenario is unlikely. Species, especially wide-ranging species adapted to or adapting to open habitats—such as dogs, lions, zebras, and the like—do not tend to show patterns of species formation within the same area. Some sort of geographic barrier or genetic isolation must accompany speciation, and find-ing such a barrier in the African Pliocene of eastern and southern Africa to account for the split between gracile and robust aus-tralopithecines is difficult to do. I considered in 1977 that the ap-parent rapid appearance of robust australopithecines in eastern Africa at about 2.3 million years ago in Omo (later redated to 2.5 million years ago), which is just north of and part of the same sed-imentary deposits as West Lake Turkana, was the result of a mi-grational event from elsewhere. The discovery of the Black Skull serves to confirm this datum point, and there is still nothing in the fossil record prior to this point in eastern or southern Africa that to me is a convincing ancestor to the robust australopithecines.

My suggestion was that the robust australopithecines had evolved in the northern African savannas in what is now the Sa-hara. Having become obligate savanna-livers, they had been sepa-rated from the savannas of eastern and southern Africa by a forest corridor that stretched from Central Africa through to the high-lands of Ethiopia. There is ample faunal and floral evidence from the site of Omo of the existence of this forest corridor from at least 4 million years ago to about 2.5 million years ago. At about this time, widespread climatic change, which we will discuss in much greater detail in the next chapter, is recorded at Omo by an increase in grassland pollen and open savanna rodents. The forest corridor opened up and the robust australopithecines spread rapidly south into eastern and southern Africa.

Until 1995 there was no fossil record of hominids in northern Africa to support this suggestion. Then Michel Brunet of the University of Poitiers and his research team reported from the 3.0-to-3.4-million-year-old site of Bahr el-Ghazal, in north-central Chad in the middle of the Sahara, an australopithecine mandible with teeth. Although the team tentatively ascribed the specimen to *Australopithecus afarensis,* the premolars are quite large and flat, and the area at the front of the jaw for the incisor teeth is quite reduced, compared to the hominids from Hadar. These are characteristics that would relate the specimen to the robust australopithecines. But the Chad mandible also has a relatively large canine and, reportedly, thin enamel on its premolars, both definitely nonrobust traits. There is too little of this fascinating fossil preserved to assess definitively whether it could be a representative of a population of hominids ancestral to the robust australopithecines, but it does establish the fact that hominids were present in northern Africa in the middle Pliocene. The preserved fauna indicates that the locality was a lakeside habitat with streams that supported some gallery forest, woodland, and open grassy areas.

The robust australopithecines were a more radical adaptation to the savanna than the gracile australopithecines—more generalized hominids that eventually evolved into the genus *Homo.* If the robusts did indeed evolve in the higher-latitude savannas of northern Africa, where climatic changes and seasonal variations would have been relatively greater than near the equatorial regions of eastern Africa, this ecological difference would help to explain the origin of their unique adaptations. Having evolved different adaptive solutions to making a living in the savanna, robusts and the descendants of the gracile australopithecines would have been able to coexist with mutually noncompeting ecological strategies throughout the late Pliocene and much of the Pleistocene. What happened to drive the robust australopithecines to extinction some 1.5 million years after their first recorded evolutionary appearance will be discussed in the next chapter.

6

The 2.8-Million-Year-Old Paleoclimatic Event Effects the Evolutionary Origin of the Genus *Homo*

Aristotle defined humans beings as the only "ethical animals." Mark Twain defined them as the "only animals that blush, or need to." Linnaeus, after classifying all the other animals in 1749, simply said *"Homo, nosce te ipsum"* ("humanity, know yourself"), but his proposed species name, *Homo sapiens* ("humanity, the wise"), betrayed his notion that it was our intelligence that set us apart from all other animals. Others have variously defined us as the only tool-using animals, the only animals that kill members of our own species, the only species with language, the only culture-bearing animal, the only animal that can contemplate its own death, the only animal that cooks its food, and even the only creature created in the image of God.

The explanations suggested to account for our appearance on earth have also been many and various. In the realm of myth and symbol, all the world's religions, whether their tenets are believed in by millions or a few dozen, have postulated a genesis for human beings. There is no single unifying theme to these theories of human origins, except perhaps that some supernatural force was responsible for human creation. In the twilight zone between myth and science, Sigmund Freud believed that men had killed their primordial father in oedipal envy to establish modern psychological human identity, at least for mid–nineteenth century Vienna. Friedrich Engels believed that humanity created itself through a conscious and Lamarckian process of "labor," thus paving the way for a socialist utopia that never arrived. Charles Darwin proposed that human beings were formed by a natural process that required no leaps of faith, no unreconstructable historical events, and no ultimately unfathomable mysteries. Darwin's theory of natural selection provided a way for science to propose and test hypotheses for human origins, and it is the basis for all that follows here.

The theory of natural selection provided no final answers on human origins. Instead, Darwin sowed a fertile field in which anthropogenic hypotheses have proliferated. Every once in a while this profusion has been pruned back. This has not always been a pretty sight—beautifully flowering hypotheses slain by the blight of ugly facts, theories only lately bloomed dying on the vine, and not infrequently much too high a quotient of manure in the fertilizer. But after nearly a century and a half of planting ideas, pruning, and weeding, a lot of the theoretical dead wood has been cleared out. What we can hypothesize and test in the realm of human evolutionary origins is now much more well constrained and controlled than at any time in the history of evolutionary biology.

No area of the anthropogenic grove has been more theoretically fertile than the evolutionary origins of our own genus, the genus *Homo*. For a long time, this ancient phase of human origins was confused with the origins of our species, *Homo sapiens,* and even with the origins of our own fully modern subspecies, *Homo sapiens sapiens.* With increasing resolution of the fossil record and better calibration of that record, we can now separate many of the steps leading to these successive stages of our evolution to modern humanity. This chapter and the next three discuss early *Homo, Homo erectus,* early *Homo sapiens,* and the dawn of modern *Homo sapiens,* all within their own ecological and temporal frameworks, as we now know them.

Brain Size and the Genus *Homo*

If bipedalism is the hallmark for Hominidae, big brain size is the sine qua non of the genus *Homo.* An earlier generation of anthropologists and anatomists thought that the increased brain size of *Homo* was an absolute, a single number of cubic centimeters within the cranium that would one day be discovered to define the genus. This value was called the "cerebral Rubicon," in reference to Caesar's crossing of the Rubicon River in his conquest of Rome. The analogy was a bit strained, but the idea was that once hominids had reached a certain critical brain size limit, they were inalterably human, and to be classified in the genus *Homo.* Modern humans have a cranial capacity of between approximately 1,500 and 2,000 cubic centimeters. Sir Arthur Keith in the 1930s thought that the correct value for early *Homo* should be 800 cubic centimeters. Others weighed in with different opinions.

There are two major problems with the concept of a cerebral Rubicon. The first problem is one of population biology, and the second is one of morphology. Natural populations of animals do not fit into neat, absolutely defined categories. Instead they vary. There will be a distribution of values for any trait, for which a mean will eventually be able to be calculated, but any single number will not adequately express this variability. We will be left with a theoretical situation in which one member of a family might be classified in the genus *Homo* and his sibling would not! Or a fossil classified as *Homo* solely on the basis of cranial capacity might be considered on all other anatomical grounds to be a good ancestor for later hominids, but a number of the later hominids might have to be classified into a different genus because their cranial capacities are below the magical number. Both scenarios do not make sense in modern population biology. Populations evolve, not individuals, and classifications need to reflect this basic fact and incorporate the intrinsic variability that populations possess.

Any measure of absolute brain size, to have any meaning biologically, has to be considered relative to overall body size. Elephants, whales, giraffes, and polar bears all have brain sizes absolutely as large as, or larger than, humans, yet their body sizes are much larger. To compare human brains to an absolute number would require that body size stayed the same throughout human evolution. We now know this did not happen. Early hominids, including the earliest members of the genus *Homo,* were significantly smaller than their descendants. A relatively small absolute size change in the brain of a hominid half our size could mean a major change in its cerebral adaptation.

Anthropologists dealt with the problem of relative brain size by calculating a ratio of brain size to body size. For a number of years this calculation was difficult because very few skulls complete enough to determine cranial capacity were found associated with elements of the postcranial skeleton, necessary to determine body size. The situation now is far from ideal but there are at least good approximations of body size and cranial capacities for most early hominid species. Using these two values, the "encephalization quotient" can be calculated. Figure 7 shows the increase in encephalization quotients through time for hominids. Instead of using a cerebral Rubicon, we can now see a definite break in the slope of relative brain size at the origin of the genus *Homo. Homo* clearly has a larger brain than earlier hominids, and its size continues to change at a rapid rate. The rapidity of brain growth in *Homo*—20 cubic centimeters of brain tissue, roughly 156 million neurons, were added every 100,000 years*—is so phenomenal that this characteristic is indeed a good bet to be the major adaptational change separating *Homo* from *Australopithecus*. We know of no other anatomical character in any part of the body that is changing so rapidly in human evolution.

Figure 7

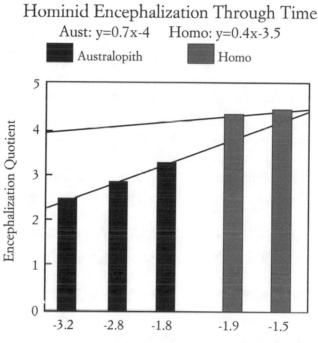

Hominid Encephalization Through Time
Aust: y=0.7x-4   Homo: y=0.4x-3.5

Hominid relative brain growth through time as measured by encephalization quotient, a ratio of brain size to reconstructed body size. The genus Homo has a higher EQ than does Australopithecus, a change that becomes apparent at ca. 2.0 ma. (Ma = millions of years ago.)

*For this calculation, *Homo habilis* (geological age = 1.9 million years) is taken at 631 cubic centimeters and *Australopithecus africanus* brain capacity (geological age = 2.8 million years) is taken at 442 cubic centimeters (see Boaz and Almquist, 1997:361). Density of neurons in the cerebral cortex is calculated by dividing estimated total number of neurons (14×10^9) by cranial capacity (1.8×10^3) in modern humans.

Brain Size, Intelligence, and Environmental Change

Intelligence may be defined as the ability to take in, process, and remember information from the environment. A relatively large brain, specifically the outer covering with all the convolutions of gray matter (the cerebral cortex), generally correlates with intelligence when compared across species of animals. Mammals tend to have bigger brains than other vertebrates. And primates tend to have bigger brains than most other mammals. Humans as a species have the largest relative brain sizes of the primates. However, at a certain level—within the species—this relationship breaks down. The brain sizes of modern people vary without too much relationship to observed or measured intelligence. For example, men tend to have larger brains than women, and even the brain of one individual will decrease in size with age. Neither fact correlates with a difference in intelligence. In other words, there is a general but not an exact relationship between size of the cerebral cortex and intelligence. When we are looking at brain size changes over spans of time 100,000 years or more in length (equivalent to 5,000 to 5,500 generations), we are probably seeing evolutionary change resulting from natural selection. We may surmise then that the genus *Homo* was undergoing severe natural selection for increased intelligence.

There have been many explanations for why humans have large brains and are more intelligent than other animals. But there is no disagreement that these are basic attributes to the species' evolutionary adaptation. People, unlike other animals, literally think their way through challenges that the environment poses. They go even a step further and anticipate problems, and they tremendously augment their own individual brain power by communicating with other members of the species in a complex interaction known as "culture." The use of stone tools has been an important benchmark for the point in human evolution that culture first appeared. We have good evidence now that cultural origins were much more of a gradual affair than at first thought. Early human culture was in some ways more similar to ape and australopithecine social behavior, and it likely evolved from these simple beginnings to culminate only much later in the cultural complexity characteristic of modern humans. Despite the fact that there was not a "cultural explosion," it is significant that the earliest big-brained fossils of the genus *Homo* and the first stone tools come onto the scene at the same time. Why should stone tool making ("culture") and increased intelligence be connected?

Anatomists interested in human evolution have observed for decades that humans have very few "built-in" physical defense, hunting, or other food procurement mechanisms. They have no sharp talons for catching prey, no strong jaws with slashing teeth for defense, no fleet running or jumping adaptations to allow rapid escape from danger, no digging adaptations to get at food or burrow underground, and no special sensory adaptations for smelling, hearing, or seeing prey.

Humans have remarkably unspecialized hand and arm structures that allow them to pick up and manipulate objects and thereby change aspects of their environment. Natural selection found it more expedient in adapting hominids to their ecological niche to change the control mechanism for these structures—the brain—than the structures themselves. For example, a human who somehow had the hands of a bonobo grafted onto his or her forearms could still carry out normal human functions because the brain would still be controlling the movements of these structures. We are all familiar with this concept, for it is not unlike marveling at how your tennis racket behaves in the hands of a pro compared to when you play, or how the old high school piano is suddenly transformed when a visiting pianist sits down to practice a few bars of Brahms Piano Concerto Number 4. The tennis pro and the pianist both have hands that in all anatomical details are the same as yours or mine, but their brains have been trained for years to coordinate very complicated maneuvers. In general terms, the human brain controls the hands and the rest of the body in the same manner, and the range of what the human brain can learn to do is stupefying.

The brain's control of the body, enabling tool use, and its increased size with greater storage capacity and integrative functions, provided early humans with a powerful mechanism of adaptation to the environment. The power of this system resides in its speed of response as well as in its flexibility. If an environmental change posed a threat to a group of early *Homo,* the group's collective brain power could be applied to the problem, and a solution found that could be learned by members of the group and put into operation immediately. In other animal species, environmental challenges were met by massive death of those individuals whose capabilities were not up to the minima imposed by the new environment. Physical changes to the new environmental demands had to wait for subsequent generations to be born. But in humans those groups with effective solutions survived the challenge, reproduced, and were consequently favored by natural selection. Natural selection caused minimal change in physical structures—the hand, for example—but strongly favored the hominids with the most intelligence. What made this system of adaptation unique was that eventually human populations confronted with environmental challenges could tolerate and cope with greater and greater change. This is indeed what happened.

DISCOVERY OF THE FIRST MEMBERS OF THE GENUS *HOMO*

Our knowledge of the first primitive humans, those early members of the genus *Homo,* is remarkably recent. *Homo habilis* was only named in 1964, and the major contributions to the fossil sample of early *Homo* were made in the 1970s and 1980s. The species is African and was the first early hominid to be age-calibrated with the newly developed potassium-argon dating technique.

Olduvai Gorge was the birthplace of *Homo habilis,* at least in terms of our knowledge of the species. Louis Leakey had been finding primitive stone tools there since the early 1930s. He was well into his third decade of looking for the maker of the tools when in 1959 his wife Mary spotted the eroding but complete skull of a robust australopithecine. Leakey at first jumped on the discovery as the reification of his belief in the existence of the Olduvai stone tool maker. But the massive cranial crests and gigantic cheek teeth of what Leakey named *Zinjanthropus* (later *Australopithecus*) *boisei* belied its reliance on extracorporeal means of processing its food. Only a year later other fossils began to turn up that showed there was a small-toothed and big-brained hominid also present at Olduvai. Enlisting the aid of South African anatomist Phillip Tobias and British primatologist John Napier, Leakey published a paper in 1964 that named a new species—*Homo habilis*—based on the Olduvai fossils. Translating as "handy" or "dexterous," the species name had been suggested to Leakey by Raymond Dart, and it aptly summarized Leakey's conception of the importance of his finds. This species had been the first stone-tool user.

In the thirty-five years since Leakey and his colleagues made this daring deduction, the evidence that *Homo habilis,* and *Homo habilis* alone, made and used the stone tools at Olduvai is still a circumstantial argument. Starting with the known facts that two hominid species were present at Olduvai and that stone tools of hominid manufacture are present there, we know that at least one of the hominids made the stone tools. Since we know that later members of the genus *Homo,* ones that lived without any coexisting robust australopithecines around, made and used stone tools, it is reasonable to assume that earliest *Homo* did as well. Similarly, since *Australopithecus boisei* had the same degree of encephalization as earlier gracile australopithecines, who did not make stone tools, it is reasonable to assume that *Australopithecus boisei* at Olduvai did not make and use stone tools either. But there is nothing in the fossil or archaeological records that unambiguously seals this argument. Indeed, some sites such as Chesowanja have

stone tools present and a fossil record only of robust australopithecines. It is possible that the robust australopithecines as well
as early *Homo* could make and use stone tools, but in my opinion
this is a less likely interpretation.

For about ten years *Homo habilis* remained recognized only
from Olduvai. Beginning in the late 1960s and early 1970s, hominid fossils began turning up in the Lake Turkana Basin of
northern Kenya and southern Ethiopia. The fossils seem to fit
into the same two categories as at Olduvai—early *Homo* and robust australopithecine. A complete skull, known by its museum
number as "1470," from Richard Leakey's site east of Lake
Turkana, was generally accepted as early *Homo,* but because it
lacked any teeth, a confident assignment to *Homo habilis,* which
had been based in part on the dentition, had to wait. Just before
his death in 1972, Louis Leakey greeted the discovery of 1470 as
proof that *Homo habilis* had been widely spread in eastern
Africa. However, this was not confirmed until Clark Howell and I
published in 1977 the description of a more fragmentary skull,
but one with almost a full set of teeth, from the lower Omo Valley, just north of Lake Turkana. We assigned this specimen, numbered L894-1, to *Homo habilis.* It was the first confirmation of
Homo habilis outside Olduvai.

WHEN DID *HOMO HABILIS* LIVE?

Olduvai Gorge was a pivotal site for paleoanthropology for another reason in addition to supplying uniquely important hominid
fossils. It was here that in 1961 Jack Evernden and Garniss Curtis
of the University of California at Berkeley first applied the then-
new potassium-argon technique for dating volcanic rocks. Many
people mistakenly think that techniques such as potassium-argon
dating provide an age on the actual fossil. In fact, the date that is
obtained comes from the rock surrounding the fossil. For this reason, a fossil's exact location within the geological strata—its

provenance—must be established beyond doubt. Geologist Richard Hay determined that the fossils of *Homo habilis* came from the lowest stratum—Bed I—at Olduvai Gorge. In 1961, Evernden and Curtis's date showed that Bed I was 1.8 million years old, an astoundingly old age at the time.

Potassium-argon dating is based on the principle of atomic decay. The ancient alchemist belief that one element could change into another turned out to be correct—one just needs a mass spectrometer to measure it. Twentieth-century physicists discovered that this change from one element to another was spontaneous, occurring at a characteristic rate of radioactive decay for each element. This knowledge was a spinoff of the research that led to the development of the atomic bomb in the 1940s. After the Chernobyl catastrophe and the widespread rejection of atomic energy, dating using radioactive elements may rank as the most beneficial peacetime legacy of the atomic theory.

Techniques that could measure the decay of one form of an element, an isotope, to another began to be developed in the 1950s. Because different isotopes decay at different rates, the dating techniques using these isotopes have different time ranges over which they can be used. Carbon 14 dating was the first of these radiometric dating techniques to be developed. It measures the amount of carbon 14 isotope, which has two more neutrons in its nucleus than the common form of carbon, carbon 12, and is formed in the upper atmosphere by natural bombardment of solar rays. Carbon 14 decays at a fast rate and is useful for determining ages in the range of a few thousand years. All of the measurable carbon 14 is gone in a sample after some 50,000 years,* so the technique cannot be applied to a problem as old as *Homo habilis.*

Potassium 40, the common form of potassium that has 40

---

*The half-life for carbon 14—that is, the time that one half of the isotope takes to decay—is 5,700 years. Each 5,700 years, then, the amount of carbon 14 in a carbon-containing sample is halved, until eventually there is just too little to measure accurately.

neutrons in its nucleus, has a much slower decay rate than carbon 14. Measurement of this isotope, which decays to argon 39, can date rocks several hundred million years old. The *upper* limit of potassium-argon dating, when there is too little decay of potassium 40 to measure, is about 1 million years ago. Evernden and Curtis chose for their analysis a type of potassium-rich rock found at Olduvai and other sites in the East African Rift Valley—volcanic rock. A volcanic ash deposit known as a *tuff* provided their sample. For each tuff sample they had to first measure the amount of potassium 40 using a mass spectrometer, a machine that hits the sample with energy and then collects the rays emitted from the sample at wavelengths characteristic for each element and isotope in the sample. The measurement of the potassium 40 in the tuff was straightforward.

It was the measurement of argon 39, a gas, that proved tricky. The decay product of potassium 40 that had built up in the tuff from Olduvai Bed I had to be collected in its entirety. If some of it had leaked out of the rock, or if some argon from the present-day atmosphere had leaked in, then the date would be wrong. Evernden and Curtis used only internal parts of the tuff, not exposed to the air surface and not affected by external weathering of the rock, for their analyses. They took particular care in the lab not to lose any of the trapped argon gas. When they had run an argon 39 test on the tuff sample, they used the value from the first mass spectrometer run of potassium 40 to determine how much of the potassium 40 had decayed to argon 39. Using the known decay rate, they then calculated the age that the volcanic tuff in Olduvai Bed I had erupted—1.8 million years ago. Because they knew from Hay's work that the fossils of *Homo habilis* had been found at the same geological level as this tuff, they concluded that this was the correct age for *Homo habilis*. Many further dating analyses in African fossil sites have confirmed Evernden and Curtis's initial findings.

The further discoveries of *Homo habilis* around Lake

Turkana extended the time period during which the species was known. Precise geological and dating work at the site of Omo by Jean de Heinzelin of the University of Gent and Frank Brown of the University of Utah established the framework. Omo contained a much thicker pack of sediments than Olduvai, with many intercalated tuffs. Omo was found to date to over 4 million years old in its earliest levels to about 1 million years old at its top. The L-894 skull was found in Omo Member G, dating to 1.9 million years ago, only slightly older than Bed I at Olduvai. Other more fragmentary fossils of *Homo habilis,* mainly teeth, were found at Omo down to Member E, dated by potassium-argon methods to 2.5 million years ago. Brown extended the Omo geological framework to East Turkana and determined that the early *Homo* fossils there were very close to the same age as the Member G. The 1470 skull was about 2 million years old and there were no fossil deposits yielding hominids between 2 and 3 million years ago at East Turkana. A fragmentary temporal bone at Chemeron near Lake Baringo, also in north central Kenya, and a mandible found at the site of Chiwondo in Malawi, both date to 2.5 million years ago, and thus confirm the date of the first appearance of early *Homo* at Omo.

As the fossil record stands now, early *Homo* appears at around 2.5 million years ago in eastern Africa. There are specimens from the South African cave site of Sterkfontein that probably represent this species as well. The earliest record of stone tools also dates to 2.5 million years ago. We have made the argument that the large human brain and the rapid and flexible adaptation to environmental change made possible by increased intelligence allowed early *Homo* to adapt to increasingly challenging environmental conditions. If this concept is correct then the changing environment in middle Pliocene Africa was the driving force behind initial evolutionary differentiation of the genus *Homo.*

ECOLOGICAL CATACLYSM ACCOMPANIES THE APPEARANCE OF *HOMO HABILIS*

Beginning in the 1960s when "ecology" became a major societal movement, integrative ecological approaches in a broad range of sciences also enjoyed their heyday. Paleoanthropology was no exception, and indeed paleoecology is an important part of the new integrative field of paleoanthropology. Field research teams interested primarily in recovering fossils and stone tools of human ancestors also began to carefully collect the fossils of *all* the animals and plants found associated with them. With these remains it would be possible to reconstruct the environmental contexts of human evolution. Some adherents of this approach went so far as to say that once you had collected the fossil or artifact and had its context established (both in terms of time and paleoecology), you could throw the specimen away and still have 90 percent of its scientific worth. No one would advocate such a heinous act; it was just a dramatic way to make the point that contextual data had become very important.

The first site at which this paleoecological approach was systematically tried out was Omo. And it was at Omo that the first firmly dated paleoecological data tying environmental change to the appearance of the genus *Homo* were gathered. Clark Howell of the University of California at Berkeley designed and directed the Omo project. Howell first established with the aid of de Heinzelin and Brown the geological framework. Within that framework his team collected plant and animal fossils that were then extensively analyzed in the lab.

Raymonde Bonnefille is a palynologist, an expert on fossil plant pollen, who works at the Centre National pour la Recherche Scientifique (CNRS) in Marseilles, France. She started working on the Omo pollen in the late 1960s when other experts said it was a fruitless pursuit. Some expert palynologists had already looked for pollen in African Plio-Pleistocene "terrestrial sediments"— deposits laid down at the edges of rivers and lakes that contained fossil remains of land-living organisms—and had found that the pollen record was very sparse. Even when there was pollen, it was so reduced compared to the diverse pollen record found in modern sediments that many experts concluded simply that what was left was meaningless. Pollen, to be preserved in abundance, needs airless conditions like the bottom of a lake or a swamp. With the periodic wetting and drying of some of the sediments at Omo and similar sites, much of the pollen had oxidized and decomposed. Bonnefille needed to search dozens of samples of sediments just to find *some* pollen, and then there were plenty of experts standing at the ready to discount her findings. But year after year she persisted, and her results were to revolutionize our understanding of the paleoclimate of the African Pliocene.

Bonnefille found that early in the Omo sequence the pollen showed a preponderance of tree species. But at the top of Member E and in Member F, about 2.5 million years ago, there was a significant shift toward a higher percentage of grass pollen in the sediments. Bonnefille boldly proclaimed that this change indicated a significant climatic shift toward drier conditions with much more savanna grassland. Despite criticisms that the Omo pollen record showed a significant destruction of pollen, no one could say how or why there should be a *differential* loss of some pollen and not others. There was no reason to believe that tree pollen would have been preferentially destroyed beginning at Members E and F, and that grass pollen would have been preferentially destroyed prior to that time. Instead it was more reasonable to expect that the same forces of destruction would have affected all the pollen in the same manner. As Bonnefille found more sites, and she analyzed more pollen, she found the same pattern. The pollen evidence showed that grasslands began to spread at about 2.5 million years ago.

Other members of the Omo research team began to find evidence that confirmed Bonnefille's findings. Roger Deschamps of the Musée de l'Afrique Centrale in Belgium analyzed the fossil wood found at Omo and discovered that the first evidence of fire scarring occurred during Member F. Naturally occurring fire is an important part of African savanna grassland ecology today. The demonstration of its presence 2.5 million years ago in thin sections of fossil wood from Omo, and the absence of fire scarring earlier in the Omo record, supported Bonnefille's conclusions.

Hank Wesselman, then one of Clark Howell's graduate students at Berkeley, studied the small mammal fossils from Omo. He also found strong support for Bonnefille's conclusions in his data. Early in the record at Omo there were more forest and woodland species—cane rats, galagoes, and even an insectivore—but beginning in Member F, Wesselman saw a change in the micromammal fauna to include a higher percentage of savanna-adapted species. The naked mole rat, a species adapted to burrowing in sandy savanna soils, showed up. The micromammals then showed the same pattern as the pollen and the fossil wood.

Other specialists on the Omo project began to look at the record for confirmatory or contradictory data on the ecological change that Bonnefille, Deschamps, and Wesselman had found. The paleontologists saw trends in the large mammal faunas. Basil Cooke noted that several of the pig lineages showed an increase in molar crown size and height, a result of adaptation to diets higher in tough silica-containing grasses. Alan Gentry saw a similar trend in the antelopes. Jean de Heinzelin and his geological colleagues found a higher percentage of fossil soils, indicating higher rainfall conditions, earlier in the Omo sequence than after Member F. A vast array of data began pouring in that in one form or another confirmed the reconstruction of a major climatic shift toward grassland savanna conditions at about 2.5 million years ago at Omo.

THE GLOBAL 2.8-MILLION-YEAR-OLD PALEOCLIMATIC EVENT

How the regional environmental history of eastern and southern Africa connected to a global pattern of climate has taken a long time to figure out. And even now there are many questions that need to be answered.

The global deep-sea oxygen isotope record shows a saw-toothed pattern of increasing climatic change beginning between 2.6 and 3.1 million years ago. The "teeth" on this double-edged saw become longer, indicating that climates became successively colder and drier. Why this should be so is a fascinating question.

From all indications, it is the onset of glacial conditions in the Northern Hemisphere that sets into motion the climatic forces that cause drier climates in Africa. In deep-sea sediments dating to between 2.7 and 2.8 million years ago researchers have found the first evidence of land sediments floated out to sea and dropped by melting icebergs. Glacial ice of the coming Pleistocene, which eventually would cover a large portion of northern Eurasia and North America, had begun to form in the Northern Hemisphere by this time. Between 2.6 and 2.8 million years ago there are increases in the amount of wind-blown dust in the deep-sea sediments off West Africa, indicating a significant increase in dry soil surfaces and a significant decrease in vegetational cover in tropical and subtropical Africa at this time. From these and other data there seems to be a good correlation between the beginning of glacial conditions in the Northern Hemisphere and dry, cool conditions in the tropics.

The ultimate cause of the temperature fluctuations in the Northern Hemisphere is astronomical. The earth's surface receives almost all of its energy from the sun. As earth wobbles through its rotation and careens around the sun in its elliptical orbit, the average amount of sunlight hitting parts of its surface varies. The sun warms up some parts of the earth, causing air to rise in high-pressure cells and winds to blow. In parts of the earth where sunlight is less intense, temperatures drop and water freezes. Paleoclimatologists call this *orbital insolation forcing* of climatic patterns. There is now good evidence that these patterns have a cycle that lasts 41,000 years in the Northern Hemisphere.

European geologists have known about the ice age and the

rock record it left behind for well over a century. What does this phenomenon of the frigid north have to do with tropical African climate, of interest to us from a standpoint of early hominid evolution? One answer to this question comes from climate models—vast computer simulations of the earth's climate. If today's wind current and surface sea temperature data are used to generate a simulation of global climate, the computer can be used to predict what changes would occur with changes in those parameters in the past. The wind flow patterns show that if the North Atlantic Ocean becomes significantly cooler, as during a glacial period, then cool dry European wind currents are pulled into and across Africa from the north and northeast (Figure 8). This prevailing wind pattern, blowing west from Africa, also opposes the eastward movement of the water-laden summer monsoons into West and Central Africa. The result is a dry and cool Africa, even if the glacial ice is thousands of miles away.

Why the Northern Hemisphere suddenly began to harbor large amounts of ice beginning at 3.1 to 2.6 million years ago, when the same pattern of orbital insolation forcing had been in action for millions of years, is a question that is still largely unanswered. Part of the answer revolves around increasing precipitation in Europe—precipitation that froze and then could not melt during the relatively cooler summers. Why precipitation increased in northern Europe at this time may be related to a stronger Gulf Stream, carrying warm water and moist air diagonally across the Atlantic from the Gulf of Mexico. Closure of the Isthmus of Panama by continental drift at about 3 million years ago, damming up westward-flowing tropical water and turning it north into the Gulf Stream, may have been the ultimate cause of the European Ice Age.

Using the deep-sea record of African wind-blown dust investigated by Peter de Menocal, we can place the first appearance of significant Plio-Pleistocene climate change in Africa at 2.8 million years ago. Because of the location of his deep-sea cores, however, de Menocal's record reflects mostly northern, western, and central

Figure 8

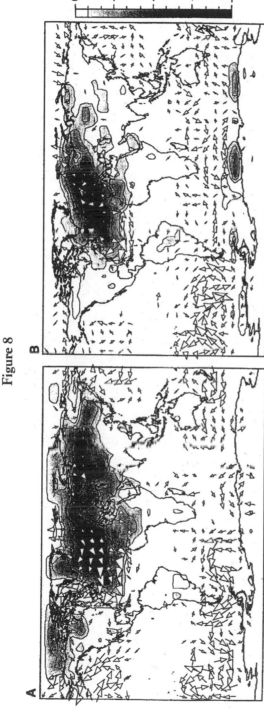

Climatic model maps that show cold winter temperatures in the Northern Hemisphere cause dry, cool northeast trade winds to blow over the Sahara and North Africa, compared to the summer monsoons when warmer and wetter conditions prevail over most of Africa. This mechanism helps to explain why glacial periods in the Northern Hemisphere correspond to arid periods in Africa and the tropics, an important reason postulated for hominid population movements in the past (from de Menocal, 1995).

Africa. Our available hominid-related fossil evidence is almost exclusively located in eastern and southern Africa, and as we have seen, it shows climate change to more open vegetation occurring some 300,000 years later, at about 2.5 million years ago. This time lag may be related to the time that natural selection took to adapt animal populations to the changing conditions. Alternatively, eastern and southern Africa may have lagged behind the environmental shift toward drier environments. With future paleontological exploration in the western and northern parts of the continent, we could expect to find earlier terrestrial indications of open savanna environments, and perhaps earlier records of the genus *Homo*, in these areas.

ECOLOGY AND GROUP DYNAMICS OF EARLY *HOMO*

For far too long, paleoanthropologists have been content simply to describe the anatomical changes that occurred in hominid evolution. Science has then stopped, at the purely descriptive phase, and speculation on the causes of the evolutionary changes has taken over. Similarly, geologists have sanguinely described and dated the enclosing and associated rocks of hominid fossils, choosing for the most part to forgo the hurly burly of paleoanthropological debate. Modern paleoanthropology, on the other hand, specifically looks for the forces of natural selection that have driven hominid evolution. Because the anatomy of the fossils, the environmental contexts in which they lived, and their exact geological ages are all critically relevant to this enterprise, everyone's data become equally important and equally subject to debate and testing. Even more subjects, particularly ecology and evolutionary biology, come into the mix at the theoretical level in anthropogeny.

Natural selection acts on individuals within populations. We desire to find and isolate those forces that acted on hominid evolution to produce the adaptations that we see reflected in the

anatomy of the fossils. What can the integrative, multidisciplinary approach contribute to an understanding of early *Homo*
evolution at this level? For example, we will need to be able to
answer a question such as, "On an average day in the summer
rainy season 1.9 million years ago, where was the individual now
known as Omo L894-1 and what was he doing at about three in
the afternoon?" Furthermore, the answer needs to be testable; in
other words, other investigators will need to be able to use similar research techniques and come up with either the same answer or a different one. All too often researchers secure in their
bastion of limited scientific certainty, be it anatomical description or geological context, have scoffed at attempts to understand early hominid adaptations in this more broadly defined
context, but understanding at this level is essential if progress is
to be made.

What do we know of the populations of *Homo habilis?* The
only way that is currently available to investigate this question is
by controlled excavation. Paleontologists Arnold Shotwell and
Michael Voorhees pioneered excavation techniques in North
America that recovered every skeletal element in the deposit and
allowed the reconstruction of population structures of extinct animals based on the numbers of specimens in each taxonomic
group. Clark Howell, my professor at Berkeley, introduced me to
this work, and I used it extensively in my excavations and analyses
at Omo. Extending an already large excavation at Omo—L398,
begun by Donald Johanson, then also one of Howell's graduate
students—I was able to obtain a number of hominids in the excavation along with some 2,500 other fossils. L398 is the earliest and
largest controlled excavation in which early hominids have been
discovered, and thus it is the first opportunity to assess early hominid population structure using this method. The site is deposited exactly in Tuff F1 of the Omo sequence, dated to 2.3
million years ago.

Nothing quite matches the tedium that accompanies identifying
in the laboratory hundreds of numbered bone fragments, all of

which have been excavated, numbered, and carefully mapped. Since every specimen must be fully identified and all the really nice specimens are readily identifiable right out of the ground, what remains for the most part needs to be identified as "unidentifiable." Sometimes I had some help in this work from Eric Meikle, one of my fellow graduate students, and to allay the boredom we would play a historical Trivial Pursuit game. I still remember reaching specimen number "1066," the Norman Conquest, "1607," the founding of Jamestown, and "2001," part of future human evolution. When this work was done, however, we could be confident that everything that was identifiable from L398 had been identified.

The bones from L398 fell into categories that paralleled the relative numbers of animals in modern African national parks. The antelopes were the most common, hippos were next, monkeys were next, pigs were next, and so on. Because the sediments of L398 were river sands, all of the bones had been moved quite a way from where the animals had lived and died, but for my purposes this was an advantage. What I had in L398 was an average estimate of African mammal populations at 2.3 million years ago, drawn from a number of habitats and over a relatively wide area. The close correspondence to modern African ecology gave me confidence that our excavated numbers faithfully reflected past population numbers.

There are 14 hominid specimens, all individual teeth and tooth fragments, from L398. The differences apparent among them and the highly sorted and transported nature of the bone sample at L398 made me think that they all derived from different individuals. More than this, 2.3 million years ago was a time of overlap of both *Homo habilis* and *Australopithecus boisei,* the East African robust australopithecine. Morphologically, some teeth can be assigned to either one or the other of these species, and the others cannot be identified more precisely than "hominid." As far as the L398 data go, the two hominid species had similar population sizes, and the assumption based on other sites where they are known to coexist is that they were about similar in abundance. Using modern population densities from African parks and the L398 excavated numbers, it was possible to get close agreements between the fossils and the modern environment. Using these numbers to calculate early hominid population densities, the L398 data showed that *Homo habilis* (along with its cousin *A. boisei*) had a very low population density, similar to that of felids (lions, sabertooths, leopards, and cheetahs). The estimated density was 0.001 to 2.5 individuals per square kilometer. This meant that on average one hominid would be seen within an area of between 0.25 and 1,000 square kilometers.

As interesting as this conclusion is in indicating that early hominids were about as rare in the environment as mammalian carnivores, it also is not a satisfactory way of expressing how hominid populations would have really been found. Higher primates are not solitary hunters but rather live in social groups. I wanted to get some idea of what group sizes *Homo habilis* may have lived in and what sizes their home ranges—the areas that groups routinely traversed in looking for food—would have been. When I published the results from L398, some theoretical ecologists provided various ideas that helped out.

Modern ecologists have discovered that the home ranges of modern mammals show a close correlation with their overall body sizes. Large mammals have larger home range sizes than do small mammals. Diet is an important factor in this relationship, however. Herbivorous mammals, such as grass-eaters (grazers) and leaf-eaters (browsers), require much less area than do meat-eaters (carnivores) and mixed feeders (omnivores), which must range over large areas to find their food. Using body weights calculated for *Homo habilis* and *Australopithecus boisei* from their skeletal bones, both species had to have been in the latter category for the L398 excavated numbers to make sense. Using the equations, I calculated that *Homo habilis* would have had a home range size of over 1,000 hectares and *Australopithecus boisei* would have had a home range size of nearly 1,300 hectares (a hectare is 0.01 square

kilometer or about 2.5 acres). An additional calculation that related body size to group size in modern mammals suggested that
Homo habilis would have had group sizes of about sixteen individuals, with *A. boisei* only slightly larger. (Modern humans using
this method have a calculated optimum group size of eighteen,
about the size of an average extended family unit.)

In the calculations of *Homo habilis* population parameters, the
numbers are approximate, and they would have varied from
group to group as well as seasonally. They show that *Homo habilis*
was by 2.3 million years ago a relatively large animal on the
African savanna, that the species was omnivorous, and that it
shared its environment with another hominid species, *A. boisei,* a
species not extremely different in its dietary adaptations. Both
species probably lived in groups averaging fifteen to twenty individuals. The population densities of early *Homo* and its robust
cousin were low, indicating that they both ranged widely over the
landscape in search of food and water.

A Synthesis

If we were now to take what is known about the patterns of late
Pliocene–early Pleistocene climatic change in Africa and put
that knowledge together with what we know about human evolution, what sort of synthesis would we have? This last section
will discuss the origin and evolution of the genus *Homo,* generally referred to as *Homo habilis,* from the standpoint of an
interplay between the paleoenvironment and early hominid populations.

The physical substrate from which *Homo habilis* evolved is
represented by *Australopithecus africanus,* a four-and-a-half-foot-
tall, 90-pound, bipedal, non-stone-tool-using hominid with a
brain size of about 500 cubic centimeters that lived around 3 million years ago. By 2.5 million years ago, five-foot-tall, 120-pound,
stone-tool-using *Homo habilis,* with a brain size of 650 cubic cen-

timeters, had evolved from a population of *A. africanus* somewhere in Africa. Over half a million years, body size increased,
stone tool use appeared, and brain size increased. Sexual dimorphism, the difference in body size and robusticity between adult
males and females, decreased from *Australopithecus* to *Homo.*

The habitats that *Homo habilis* lived in were part of the
African savanna, broadly defined. We can deduce from the excavation evidence that groups covered a large amount of territory in
the course of a year's foraging for food and water. Habitats needed
to include a predictable source of drinkable water since the physiology of apes and modern humans both still require large amounts
of water. Humans, probably beginning with *Homo habilis,* also
sweat and need additional water for evaporative cooling. This intriguing paradox both frees hominids from water for relatively
short periods of time and ties them ever more closely to sources of
fresh water. Lakes, rivers, and streams on the African savanna are
surrounded by gallery forests and woodlands, indicating that hominid habitats would have included trees.

The ecological relationship between early hominids and trees
has been the subject of paleoanthropological speculation for some
years now. Beginning with anatomical evidence from hand and
foot bones that australopithecines may have climbed and hung in
trees, others have used environmental evidence to deduce that
early hominid habitats may have been in the more wooded portions of the savanna. The earliest hominids, *Ardipithecus ramidis*
and perhaps *Australopithecus anamensis,* may have actually lived
in what we would term more accurately forest. New pollen evidence from the South African *A. africanus* sites of Sterkfontein
and Makapansgat indicate that they had a higher component of
trees around than previously thought. And even at the *Homo habilis* stage, new oxygen isotope analyses of soil sediments from
Olduvai Gorge Bed I by Nancy Sikes indicate that the hominid
habitats would have been wooded. Trees may have been important food sources for hominids, who relished fruit, but they may
have also been important refuges from predators. Small trees,

particularly acacias, would have been important shady places on the open savanna.

Darwin considered the African savanna one of the "most dangerous places on earth," with its numerous large mammalian predators. Placing early hominids in this environment has always seemed problematical. How did they survive without shelter, great speed, natural defenses, and weapons? Trees may have provided some of the shelter since such predators as lions, hyenas, wild dogs, and cheetahs do not climb. But for relatively large, group-living primates, long-term residence in trees, including sleeping there, would not be a very good solution. The most significant predator, the leopard, which left its tooth marks on more than one early hominid skull, habitually climbs and lounges in trees, and it also hunts at night. This fact alone would make any *Homo habilis* parent worry about the safety of the children at night. *Homo habilis* consequently must have constructed "shelters," which probably were no more than ape or australopithecine "nests" with walls and a roof to keep out the predators. Because *Homo habilis* body size was relatively large (like australopithecine, chimp, and gorilla body sizes), these were constructed on the ground. At site DK at Olduvai, a circle of stones associated with many stone tools may be the first remnant in the archaeological record of one of the shelters, with the stones serving to anchor the thorn tree branches that probably formed the walls and roof.

Considering *Homo habilis* defenseless is probably a serious underestimate of its capabilities. As a setup for *National Geographic,* Louis and Richard Leakey once tried to ward off a pack of hyenas in Kenya from a fresh kill with nothing but branches and a lot of yelling. Theatricality aside, this experiment showed that hominids without modern weapons, fire, or even stone tools could mix it up with large mammalian predators on the savanna and live to tell about it. The hyenas got most of the meat, however. Direct confrontation—hand-to-hand combat—was certainly not the hominids' strong suit.

Early hominids had several unique defenses against direct predation and resources to draw on to compete for food with other mammals on the savanna. Like chimps and modern humans, they were loud. Generally, only animals that are ecologically dominant such as lions and elephants make a lot of noise, and when they do, it is a very effective reminder to lesser animals to steer clear. Hominids mimicked this existing aspect of interspecies competition and scared away many animals just by making a lot of noise. This method still works on the African savanna today. Also, early hominids, like apes and modern humans, could strike out at long distances by throwing. Eileen O'Brien of the University of Georgia has made the argument that aimed throwing is a potent defense mechanism, and it is indeed astounding to see the accuracy and power with which simple rock projectiles are delivered by some human groups in recent recorded history. A stone between the eyes to an advancing leopard would certainly slow him down, an effect that could be augmented by flanking adult male hominids screeching and brandishing acacia branches if more dissuasion were needed.

If all else failed, hominids could run for it. If they waited too long, though, they were done for. Just about every other mammal on the savanna can run faster than hominids in the short-distance events. But hominids excel in the long-distance races. If they have a long enough head start, which they would get from a high vantage point and an alertness about what was happening around them, they could outdistance lions, hyenas, and even hunting dogs if they needed to.

What *Homo habilis* ate can be gauged in general terms from its population densities and its overall habitat. The numbers of hominid fossils found in controlled excavation are too small for a savanna-living, herbivorous, and water-tied species. Like their modern counterparts, early hominids were omnivores, and they foraged over large home ranges. We can be relatively certain that early hominids ate a good proportion of the native African fruits and plants that chimps and modern Africans eat, which do not require preparation by fire or by some involved method of detoxification. Figs, berries of *Grewia* bushes, certain grass seeds and shoots, and edible roots were eaten. Bananas, which were brought into Africa from southeastern Asia, and manioc, which was also brought in from Asia and whose root requires extensive preparation by modern Africans before it can be eaten, were not part of early hominids' diet. We can imagine the effect on the vegetable component of early hominids' diet during arid climatic phases and droughts when we reflect on the number of succulents and high-rainfall plants in our modern diet. Lettuce, pineapple, mango, oranges, spinach, grapes, bean sprouts, cabbage, peas, and tomatoes all require high amounts of water. Plant species like these are rare in open savanna habitats in Africa today, and it is unclear where they would have grown during *Homo habilis* times. Although it is certain that *Homo habilis* subsisted on many plants, we have much research to do before we know for certain what they were and how they would have been affected by climate change.

Unlike the plant diet of *Homo habilis,* we do have actual evidence of what animal remains were eaten. The animal protein part of *Homo habilis* diet can be partially reconstructed from the cut marks left by their stone tools on animal bones. It is no surprise that some of the most popular animals seem to have been tortoises. At Olduvai and at the slightly older site of Senga 5A in eastern Zaire, land tortoises of the genus *Pelusios* were eaten. These animals were one of the few that were slower and smaller than hominids, and hominids were able to breech their defense mechanisms. The stone tools that *Homo habilis* possessed allowed them to break open and get inside tortoises' shells, whose round contours rendered them impervious to the powerful jaws and teeth of lions and hyenas. The tortoises' habit of burrowing into the ground for protection did not save them from an animal that could dig them out with a stick. In addition to tortoises, the relatively defenseless old and young of larger animals were probably the preferred game of *Homo habilis* "hunters." In fact, much of the meat that *Homo habilis* groups got their hands on may have been leftovers, scavenged from the kills of the big cats, hunting dogs, or hyenas. Some archaeologists believe that the "overprinting" of carnivore bite marks by hominid stone-tool cut marks on the bone remains of big mammals like elephants, hippos, and large antelopes indicates that hominids arrived at the scene late and were essentially scavengers.

Our picture of *Homo habilis* is of a small, resourceful hominid that was able to protect itself and forage effectively in its savanna-woodland habitat. The social group to which each hominid belonged was critically important to its survival. The environment was a changing mosaic of vegetation types, water resources, safe places of refuge, and sources of animal protein. How to get at these resources was a daily challenge met by accurate assessment of the environment by the members of the group. Natural selection was a hard and heartless taskmaster. If hominid groups made the wrong assessments, they paid for it by death from thirst, weakness, or death from starvation, increased predation, or increased chance of death or injury from trauma. Since hominids adapted to the environment primarily by their wits, there was a tremendous pressure for increase in intelligence. We see this reflected in the increasing brain size in *Homo habilis* and its descendants.

Homo habilis was a successful member of its community, along with its hominid cousin, the robust australopithecine. This coexistence of hominids lasted for approximately 1 million years, during which time ecological conditions stayed variably the same. In other words, the fluctuations in climate to which hominids had become adapted stayed within the limits of their adaptations. There were periodic climatically arid periods occurring every 21,000 years, followed by phases of warmer and wetter climates, but through it all hominid adaptations had enough flexibility "built in" to weather the changes. This pattern began to change gradually as the Pleistocene epoch approached. Change was greater and environments in Africa were harsher than anything that had occurred before. Natural selection transformed *Homo habilis* into a new and very different hominid, at the same time that changing environments so drastically altered conditions that the robust australopithecines could no longer survive. We will now turn to the twists and turns of human evolution in the Pleistocene and how ecological history shaped the origins of our own species.

7

A "Paleoclimatic Pump" Expels *Homo erectus* out of Africa

When Dutch anthropologist–physician Eugene Dubois discovered in the 1890s the fossil skullcap and thighbone of a new form of human being at the site of Trinil in Java, he termed it *Pithecanthropus* ("ape–human"), using a term previously coined by the German naturalist Ernst Haeckel. Haeckel had suggested also a species name, *alalus* ("without language"), to go along with his new hypothetical genus name, but in officially applying *Pithecanthropus* to real fossils, Dubois ignored it. Instead he used a term to underline what he thought was the most important defining characteristic of the most primitive hominid to have been discovered up to that time—*erectus*. Dubois deduced from the straight unapelike shaft of the femur that *Pithecanthropus* had indeed walked erect. *Pithecanthropus erectus* then had shared erect posture with modern humans despite the heavy browridges over its eye sockets and other characteristics of the skull that Dubois interpreted as apelike.

Generations of paleoanthropologists were to pick apart Dubois's theoretical formulations. A better appreciation of the natural variability of animals within populations and discoveries of many more fossil hominids made it very unlikely that the specimens from Java were in actuality a different genus from modern human beings. Most specialists decided to sink *Pithecanthropus* into our own genus, *Homo*. Thus, Dubois's fossils became known as *Homo erectus*. Also, discoveries in Africa in the early to mid–twentieth century demonstrated that bipedalism had occurred far earlier in hominid evolution than the *Homo erectus* stage. What Dubois had considered perhaps most significant about his discoveries had in fact been inherited from australopithecines and his species name turned out to be a misnomer. But, misnomer or not, *erectus* is the first zoological name proposed for fossil specimens now recognized as a distinct fossil species, and thus by the rules of zoological nomenclature, the correct designation for Dubois's fossils is *Homo erectus*.

After the initial discovery of the species in Asia, *Homo erectus* fossils from Africa that were both more complete and very much older eclipsed those from Java. It was widely accepted by paleoanthropologists that *Homo erectus* did not migrate into Eurasia from Africa until about a million years ago. New discoveries within the past five years, however, have given back to Asia some of the limelight. It is now apparent that *Homo erectus* appeared in the fossil record of Asia at 1.8 million years ago, the same time as its first appearance in Africa. We shall see that this "early" appearance of *Homo erectus* in Asia was caused by climatic change at the dawn of the Pleistocene. Changes in climate that profoundly affected fauna, flora, and habitats pushed hominids out of Africa, their ancestral home, and set them on a course for populating the rest of the Old World.

THE BEGINNINGS OF *HOMO ERECTUS*

Although Java, a large island now part of Indonesia, was the first locus of discovery of *Homo erectus,* it is obvious that the species would have originated from a mainland Asian species. Even back in the nineteenth century it was widely appreciated that "land bridges" had to be invoked to explain the occurrences of animal and plant species on islands. Java sits atop a relatively shallow continental shelf known as the Sunda Shelf that during low stands of the Pacific would have connected it to mainland Southeast Asia. The mechanism of this periodic lowering of sea level throughout the Pleistocene was tremendous freezing of water on land by the expanding glaciers of the ice ages.

The expected discovery of the mainland hominids that had walked across to Java during the Pleistocene happened first in China, beginning in the 1920s at the famous site of Zhoukoudian (originally spelled Choukoutien). Although these fossils were also first named a new genus *(Sinanthropus),* according to the fashion of the times, they too were later recognized to be members in good standing of the species *Homo erectus.* The Zhoukoudian site was vastly more productive than the Javan sites and yielded the largest single population of fossil hominids (estimated to be forty individuals) yet found anywhere. The skull form seen at Zhoukoudian was similar to that of the Javan *Homo erectus.* This was sufficient reason for most paleoanthropologists to believe that the Asian stock from which Dubois's hominid had sprung had been found, despite the fact that Zhoukoudian dated to around 500,000 years ago, more than a quarter of a million years younger than Trinil.

After World War II and the irretrievable loss of the Zhoukoudian fossils during their shipment out of China in 1941, a few specimens discovered in Europe were put forward as representative of this stage of human evolution. The most well known of these is a jawbone from near Heidelberg, Germany, dated to about 500,000 years ago and known as the Mauer Jaw. Like *Homo erectus,* it has a receding chin and a massive ramus connecting up to the jaw joint. Most experts now consider it an early example of *Homo sapiens,* although others accept it as the type specimen of a unique species intermediate between *erectus* and *sapiens* known as *Homo heidelbergensis.* No fossil specimens found in Europe can be attributed with any certainty to *Homo erectus.* It was instead Africa that provided new discoveries of this stage of human evolution.

As with so many other groundbreaking paleoanthropological discoveries, Louis Leakey was the source of the first fossil remains of *Homo erectus* from Africa. At Olduvai Gorge Bed II he found a skullcap, missing the face below the brows and the base of the skull, which was catalogued as "Olduvai Hominid 9." The specimen shows the unmistakable browridge and cranial vault anatomy of Dubois's species. Although it was originally called by various names including *Homo leakeyi,* it is now generally accepted as African *Homo erectus,* with an accepted date of about 1.2 million years ago.

Other discoveries at Swartkrans in South Africa and around Lake Turkana, Kenya, confirmed the ancient presence of the species in Africa. A nearly complete skeleton of a twelve-year-old boy west of Lake Turkana at a site called Nariokotome is the best single specimen of *Homo erectus* known. The accurate dating record in eastern Africa showed that *Homo erectus* was much earlier here than at the Javan and Chinese sites—roughly 750,000 years ago for Trinil and 500,000 years ago for Zhoukoudian. In Africa the species was clearly present by 1.6 million years ago, the date of Nariokotome, and its earliest appearance is now documented by fragmentary finds around the Lake Turkana Basin to 1.8 million years ago.

On the basis of anatomical evidence, particularly of the cranium, *Homo erectus* seems to have evolved from *Homo habilis.* Specimens, particularly from East Lake Turkana and dating back to more than 2 million years ago, show transitional traits between the two species. Some specialists have given these pre-*erectus* fossils a new species name—*Homo ergaster*—but most still consider them part of *Homo habilis*. In the transition from *habilis* to *erectus*, the browridges enlarge, the face becomes smaller while the part of the skull holding the brain becomes larger, the overall shape of the skull grows to be longer and flatter, and the teeth become somewhat smaller in size.

NEW DATES AND A NEW SCENARIO FOR ASIA

During the 1990s new dating of the earliest *Homo erectus* sites in Java and a new discovery from the Georgian Republic have revolutionized our concept of the earliest beginnings and geographical spread of *Homo erectus*. We now know that hominids were part of the new Eurasian fauna ushering in the Pleistocene epoch some 1.8 million years ago.

The discovery of a very early date for hominids in Java goes back to 1971. In that year Berkeley geochronologist Garniss Curtis and Indonesian paleoanthropologist Teuku Jacob reported a potassium-argon date run in Curtis's lab of 1.9 ± 0.5 million years ago on a piece of volcanic pumice from a site near Perning, Java. The earliest hominid is a child's cranium found at Mojokerto, Java, by an Indonesian collector by the name of Tjokrohandojo in 1936. The pumice that Curtis dated was thought to derive from the same stratigraphic level, but there was some uncertainty. Most of the problem with the date came from the big "plus–minus" error attached to the end of it. This is a statistical abstraction of uncertainty that geophysicists used to estimate the variability between analytical runs in the mass spectrometer. The relatively large uncertainty associated with the Perning date had to do with the sample being relatively small and perhaps contaminated by some degree of weathering of the rock itself. Most paleoanthropologists ignored the date or relegated it to a footnote because of these uncertainties.

Two researchers who did not ignore Curtis's date on the Perning pumice were geologists Dragan Ninkovich and Lloyd Burckle of Lamont–Doherty Earth Observatory. In papers published in 1978 and 1982 these researchers used the dating and paleoclimatic record of the deep-sea core to tie in the hominid-bearing strata of Java. Ninkovich, a vulcanologist, was able to correlate a layer of tektites—small tear-shaped pieces of glass probably resulting from meteorite impacts with the earth's crust—from the fossil sediments on Java to a layer of microtektites found in the deep-sea core. Their dating constrained the age of the top of the *Homo erectus*–bearing fossil sediments to about 700,000 years ago. Burckle, a specialist on fossil diatoms—small marine plants with hard shells—was able to match up the sediments underlying the hominid fossil beds in Java with the deep-sea core on the basis of the diatom species they both contained. His work provided a lower limit to the Javan hominid dates. He concluded that the base of the hominid-bearing strata in Java was between about 1.9 and 2.1 million years ago. Curtis's date fit in perfectly with these results, and Ninkovich and Burckle went so far as to reconstruct the paleogeography of the land connections between Java and mainland Asia and to review when the first land mammals migrated across the land bridge.

At about this same time, researchers in China came up with a date of 1.7 million years ago for some early hominid teeth found at the site of Yuanmo in southern China. They based their age assessment on the record of paleomagnetic reversals in the earth's magnetic field preserved at the site. Western researchers again were skeptical and one paleoanthropologist, Geoffrey Pope, was only willing to accept the presence of hominids in Asia by about 1 million years ago. Doubt centered around both the geological contexts of the original fossil finds and the technical methods used to date the sediments.

There matters stood until, in 1994, Carl Swisher, Garniss Curtis, and colleagues at the Berkeley Geochronology Center redated volcanic sediments in Java to just over 1.8 million years old. Technically flawless, their method consisted of extracting minerals from the original Mojokerto child's cranium found in 1936 and matching them chemically to sediments at the reported discovery site. They matched. The researchers then proceeded to date the sediments at the site using an advanced ^{40}Ar-^{39}Ar technique. They obtained a very accurate date of 1.81 ± 0.04 million years ago, thus vindicating Curtis's date twenty-three years earlier. An archaeologist commentator in the British journal *Nature,** forgetting or ignoring Ninkovich and Burckle's work, wrote that, "the dates . . . have to be taken at face value because they cannot be cross-checked with any closely associated biostratigraphic data." Ironically, the fact that the date by Swisher and colleagues correlates so well with all the available biostratigraphic and contextual evidence is exactly why it is likely to be correct.

Ninkovich and Burckle pointed out in 1984 that the island of Java likely dates to the age of a deposit of volcanic ash in western Java exploded subaerially—that is, on land—at 7 million years ago. The earliest known fossil mammals on Java, however, correlate with species characteristic of the Indo-Pakistan Pinjor fauna, with a maximum age of 2.4 to 3.0 million years ago. The clear implication is that Java was formed but stayed unpopulated by large mammals for some 4.0 to 4.5 million years. Then at the initiation of large-scale global cooling and glaciation at the end of the Pliocene, sea levels were lowered sufficiently to establish a connection of dry land between mainland Asia and Java, via Malaysia and Sumatra (Figure 9). Large mammals walked across this land bridge and left their fossils at several sites in western Java. Hominids are not known from these western Javan sites that are stratigraphically below the hominid-bearing sites of eastern Java. Neither are hominids known from the similarly aged Pinjor faunas of India and Pakistan, which have been investigated by paleontologists for well over a century, or from any other of a number of well-investigated late Pliocene fossil sites in Eurasia. Hominids may turn up one day at one of these sites, but

*Gamble, Clive, 1994, Time for Boxgrove man. *Nature* 369:275.

Figure 9

Southeast Australasia during a low stand of sea level during a Plio-Pleistocene glacial period, showing the contiguous landmass of "Sundaland," a name derived from the Sunda Shelf that now is under water. Early hominids (Homo erectus) dispersed to Java by 1.8 ma but not to Australia, across Wallace's Line, until the latest Pleistocene (from Boaz and Almquist, 1997). (Ma = millions of years ago.)

on present evidence it seems a fairly safe bet to presume that they had not yet migrated out of Africa.

A single mandible of a probable *Homo erectus* was found by paleontologist Leo Gabunia at the site of Dmanisi, Georgia, in 1994 and dated by both potassium-argon and fauna to 1.8 million years old. Most western paleoanthropologists are not conversant with the Plio-Pleistocene faunas of Central Asia and thus were not sure how to assess the context for so early a date for hominids in mainland Asia. Gabunia presented his arguments for the accuracy of the date at a research conference held at the International Institute for Human Evolutionary Research in Oregon in 1995. His arguments that the Dmanisi hominid mandible is a part of a fauna that dates to nearly 1.8 million years ago was convincing to a majority of the paleontologists at the meeting. Except for the near-contemporaneity of some early teeth from China, Dmanisi represents the oldest known record of hominids in mainland Eurasia.

Since the mid-1970s our conceptions of the age and geographical spread of *Homo erectus* have changed dramatically. First, Africa was discovered as the continent of origin of the species. Second, this origin was found to be quite ancient, extending back to fossils 2 million years old. Third, discoveries in only the last few years have shown that *Homo erectus* was present in Asia at ages as old as the African record. The only reason that these discoveries have not replaced Asia as the birthplace of *Homo erectus* is that there is no prior record of hominids in Asia from which *Homo erectus* could have evolved.

The beginnings of *Homo erectus* in Africa and its spread to the rest of the Old World are momentous occurrences in human evolution. What forces mediated the appearance of *Homo erectus* and what caused the migration of the species out of Africa (and perhaps back and forth throughout the Old World) are the subjects of the next sections.

HOMO ERECTUS AND THE BEGINNINGS OF THE HUMAN EXPERIMENT

Although *Homo habilis* is the earliest generally recognized member of the genus *Homo,* it is with *Homo erectus* that "humanness" truly begins. These hominids first started to make sophisticated stone tools requiring more than two or three flakes knocked off to produce a simple cutting edge. They first began to use fire. They spread out farther in their range than any other modern primate except *Homo sapiens.* They were probably the first hominid species to exclude all others in the ecological competition for food and space, likely driving the last of the robust australopithecines in Africa to extinction. *Homo erectus* was the first hominid able systematically to hunt and kill animals larger than itself, as indicated by faunal remains from several archaeological sites. Part of its prey seems to have been other *Homo erectus,* whose skulls were broken to extract their brains and whose limbs were roasted. This last deduction about the behavior of *Homo erectus* has been controversial since it was first suggested by Franz Weidenreich in 1939 on the basis of the fossils from Zhoukoudian. But there seems little to contradict this conclusion except a misplaced Rousseauian belief in the primeval "goodness" of humanity.

A hominid of relatively large size, *Homo erectus* weighed in at 100 to 150 pounds. Its body in most measurable estimates was fully human and essentially the same as ours. Two intriguing differences from *Homo sapiens,* however, are found on the Narioko-tome skeleton. *Homo erectus* possessed a substantially smaller spinal cord, presumably because the species' brain size was significantly smaller than ours. The rib cage of the species also showed the conical, apelike shape of the australopithecines, a surprisingly primitive characteristic for such an otherwise advanced member of the genus *Homo.*

It is the skull of *Homo erectus* that was significantly different from *Homo sapiens,* indicating not only that what went on inside was qualitatively dissimilar from what goes on within our neural circuitry but also that the head was used in a different way. The brain was small by modern human standards, averaging from 900 to a little over 1,000 cubic centimeters. Microcephalics—people whose brains are congenitally small or whose skull bones fuse prematurely—today have brain sizes this small, but these individuals usually suffer severe mental impairment. *Homo erectus,* when fully adult, probably had the mental abilities of an average five-year-old. It has been suggested that *Homo erectus* did not learn to talk until adolescence. Whatever specific behavioral adaptations *Homo erectus* had, "unhuman" though they may seem by modern standards, they certainly constituted a successful adaptation at the time. It is probably a secure deduction that *Homo erectus* depended on fairly standardized behavioral solutions to environmental problems. These were perhaps "ritualized" in a way that modern humans would find unimaginably tedious—analogous to an adult's reaction at listening to the hundredth rendition of the ABCs by a five-year-old on a rainy afternoon. The five-year-old eventually grows out of this stage. *Homo erectus* individuals did not.

The skull bones of *Homo erectus* are inordinately thick. The browridges project over the orbits, protecting them with a dense shelf of bone. The back of the skull projects backward into a "torus" of bone that overhangs the back and sides of the neck. The form of the skull itself is turtlelike, low and arched up to a ridge of bone running the length of the skull. This peculiar morphology has baffled anatomists. The only suggestion that seems to make any sense is that the skull of *Homo erectus* was adapted for protection from external blows, from the front and back as well as from above. What specific forces of natural selection caused this morphology to evolve is still speculative. There is nothing in the underlying shape of the brain (that in any event forms before the skull bones in development) to suggest this sort of morphology. There is nothing in the transmission of chewing forces causing buildup of supporting bone to explain this skull shape. My own best guess is that it had to do with protection.

All the morphology of the *Homo erectus* skull vault can be explained as a design to withstand force administered from the top and sides. Was this anatomy a "hard-hat" adaptation occasioned by living in cave environments with low ceilings and falling rocks, or a "helmet" adaptation resulting from ritualized conflict within the species that perhaps involved striking the head with sticks or clubs? Unfortunately we do not know. We might tend to dismiss the idea of *Homo erectus* men (and women) systematically bashing each other over the head, but there are modern analogues and there are several additional aspects of the behavior of *Homo erectus* that we should consider.

The available archaeological data on *Homo erectus* reveals that one type of stone tool was used for about a million years—*one* type of stone tool, for a *million* years, all over Africa wherever *Homo erectus* is found after 1.4 million years ago. For some reason it is not associated with Asian *Homo erectus.* This stone tool is the Acheulean hand ax. It is not an easy tool to make and modern *Homo sapiens* graduate students are not able to fashion a very good one even after an entire academic term of practical experience. The implication is that *Homo erectus* would have expended a tremendous amount of time and energy—years—laboriously learning how to make hand axes. The technique must have been passed on by rote repetition. Hand axes stayed the same for untold generations.

This method of transmission of cultural knowledge is totally foreign to us. Nothing that we *Homo sapiens* learn and internalize stays the same. We have to change it, improve it, make it look better, modify it to fit our specific needs—be it a chair, an art form, or our own language. But this never occurred to *Homo erectus,* not in a million years. The behavior of *Homo erectus,* despite its complexity compared to earlier hominids and apes, must have been quite formulaic and ritualistic compared to our own. There is no evidence of *Homo erectus* art, music, carving, or elaboration of any cultural attributes over the necessary and utilitarian. *Homo erectus* did not bury their dead and thus probably did not have any form of belief system approximating human religion, despite the inferred ritualistic character of their behavior. If *Homo erectus* had a way of dealing with internal group conflict as well as perhaps intergroup conflict, it likely stayed invariant for so long that morphological evolution would reflect the behavior. A modern analogue may be the Australian aboriginals who have the thickest skulls of any modern people and who are known to have had the widespread practice of hitting each over the head very forcefully with sticks to settle disputes. Usually the wielders of the sticks are men and they are usually fighting over one of two things—women or land. The blows administered can be fatal. It may have been the same for *Homo erectus*.

Despite the simplicity of the culture of *Homo erectus,* there is one cultural possession that the species had that was transcendentally important. It constitutes the primary cultural patrimony that *Homo erectus* bequeathed to us. It is fire, that strange oxidative reaction that yields heat, light, and smoke. It can magically transform food from raw to cooked and, when unleashed, it can become a force of environmental change of immense power and destruction. We still stare at it in dumfounded awe when it is in the form of a simple campfire. Its various uses in the modern world range from gas ranges, internal combustion engines, and Bic lighters to the derivative microwave ovens, nuclear reactors, and electric lights. Where did it come from and how did *Homo erectus* harness its power?

HOMO ERECTUS AND FIRE

Fire is the most powerful tool of adaptation and change that humans possess. Its origin as part of the human adaptation must be ecological, prompted by a need to survive and find food in an environment that somehow was too challenging for the old ways of *Homo habilis.* For many years the standard paleoanthropological belief was that fire began in the cold Northern Hemisphere during the impending ice age, yielding heat and light for cave-dwelling hominids. But evidence has now been found to indicate that early *Homo erectus* in tropical Africa, not Eurasia, first learned to control and use fire.

J. W. K. (Jack) Harris, an archaeologist at Rutgers University, has discovered at the Kenyan sites of Chesowanja and East Lake Turkana areas of baked clay that date back to about 1.5 million years ago. These circular areas of baked clay test out to have been formed by quite limited fires of low temperature, unlike widespread spontaneous bush fires or high-temperature lightening strikes. Their circular outlines are different from the pattern of soil baking that results from a burning tree stump. At the South African site of Swartkrans, Andrew Sillen and Bob Brain have reported convincing evidence of fire-cracked bone dating back to 1.2 million years ago. None of these sites have "hearths"—the stone-rimmed, ash-filled, and carbon-rich deposits indicative of habitual fire use—that archaeologists have discovered at later time horizons around the world. Consequently the argument for the early evidence for the controlled use of fire by hominids has been controversial and largely unaccepted by the community of archaeologists. The old association of glacially cold conditions and fire has also died hard. How does it make sense that fire use may have originated in the relatively warm regions of the earth near the equator?

Since many past generations of paleoanthropologists have been resident in the Northern Hemisphere, it is understandable that their primary association with fire is warmth and light. But it is more likely from an ecological standpoint that the initial importance of fire to early hominids was its destructive power as used in the pursuit of food. As scavenger hunters, hominids would have been drawn to bush fires because animals are pushed before the advancing flames, and small or immature animals are sometimes killed and partially roasted. An opportunistic ecological relationship of some sort certainly existed between savanna bush fires and hominids back to early australopithecines, long before there were any hominids outside the African continent. Naturally occurring bush fires are common in the African savanna today and they are important to maintaining the savanna ecosystem. They are caused by lightening strikes and during the dry months of the year they burn the grass over large areas of savanna. They burn fast and leave almost all the standing trees. Occurring most commonly at the end of the dry season, they clear the ground for new grass shoots when the rains begin. At night they light up the sky with an eerie red glow. Early hominids would have noticed them, perhaps at first to avoid being caught in their path, and later for their food potential.

The first ecological association between hominids and fire then was almost certainly a food relationship. If fire was curated—that is, kept around for use when needed to set part of the savanna on fire to drive out small animals for food—then hominids would have become accustomed to its properties and its potential uses. The abundance of naturally occurring fire would have been greatest during the dry season, the same time of year that food availability would have been the lowest, and thus it is likely that fire use would have started out as a seasonal activity. This association between fire use and the dry season is key to understanding the increasing importance of fire in the adaptations of *Homo erectus*.

The "Dog Days" of the Pleistocene and the Rise of *Homo erectus*

Homo erectus begins its evolutionary career at the beginning of the Pleistocene. The Pleistocene was the last of the epochs of geological time before the Recent period, beginning at about 1.8 million years ago and ending, more or less by convention, at 10,000 years ago. The Pleistocene is typified by an increasing amplitude of environmental change. The dry, cold periods become drier and colder, and the wet, warm periods become wetter and warmer. Having ancestrally come from warm and wet habitats, *Homo erectus* was equipped to deal with these changes. We modern *Homo sapiens* still go on vacation to the warm and wet tropics—to lie on the beach, just so long as we have plenty to drink and shady palm trees—but going to extremely arid environments such as the Sahara or the moon require major expeditions. The periodic swings to aridity during the Pleistocene posed environmental challenges of equal magnitude to *Homo erectus.*

The deep-sea oxygen isotope curve demonstrates how significant these environmental changes were. Beginning at about 1.8 million years ago there was a significant increase in wind-blown dust off the West African coast, indicating increased aridity. Fluctuations in average sea surface temperatures were significant, ranging over 10 to 12 degrees Centigrade (about 18 degrees Fahrenheit). On land, without the ameliorating effects of the sea, extremes could have been even greater. Oxygen isotopes in carbonate deposits in East African soils show a marked increase in savanna vegetation at 1.8 to 1.6 million years ago. We can think of this change as the difference between a relatively pleasant summertime temperature of 78 degrees Fahrenheit with moderate humidity and a "heat wave" of 96 degrees with low humidity. In the vernacular of the American South, the very hot and dry weather of late summer is termed the "Dog Days." In the days before air conditioning, this period of the year was dreaded as unpleasantly hot. For *Homo erectus,* conditions of extremely hot and dry climate were much more serious.

When there was no rain, there was no food for *Homo erectus* to eat. The grass dried up, the animals left, there was almost no fruit on the trees, and shoots and food plants failed to grow. Sometimes there was not even enough water to drink. Today when there is drought in Africa there is mass starvation and death. The *Homo erectus* death toll that natural selection exacted during periods of climatic aridity must have been tremendous. A few straggling groups managed to survive, however. Intelligence, culture, group cooperation, tools, and fire were probably their survival tools.

Migration is another way that animals adapt to changing envi-

ronmental conditions. Sometimes this migration is an annual affair, following the seasonal availability of water and forage, as in
the case of the wildebeests of the Serengeti Plains, or avoiding the
onset of conditions inimitable to survival, such as migratory birds
flying toward lower latitudes to avoid the onset of the freezing
temperatures of winter. Mary Leakey and two ecologists have actually suggested that this sort of seasonal migration may have typified early hominids in eastern Africa. They suggest that as the
herds moved out of the Serengeti in the summer dry season, hominids may have followed them. As support, they note that no
carnivore alive today migrates with the herds and that not only
meat but more abundant plant resources would have been available to a migratory band of hominids. As the winter rains returned, so would the grass, the herds, and presumably the
hominids.

Regardless of the correctness of this scenario, it is certain from
an ecological standpoint that early hominids would have spread
out within their home range consistent with the seasonal presence
of fruiting trees, other sources of plant foods, and sources of meat
or protein. If, because of lack of rainfall or other environmental
factors, their normal home range did not have the necessary food
or water resources to sustain them, they would have been forced
to travel farther afield to find them. There are two primary ecological consequences of this fundamental change in food resource
availability. First, home range sizes would have increased in response to greater aridity in the environment, leading to range
overlap and competition with neighboring hominid groups and
other species competing for the same resources. Second, greater
migration would ultimately lead to greater dispersal—that is, the
result of many small migratory events would have tended to push
hominids into regions that they had not previously inhabited, if
these were habitats that could be life-sustaining. Both of these deductions explain two of the most significant events in the Pleistocene history of hominid evolution: the spread of hominids out
of Africa and the extinction of the robust australopithecines.

LEAVING THE MOTHER CONTINENT

Africa, the birthplace of the hominid lineage, is a huge continent with many diverse habitats. It had been cut off for varying periods of time from the rest of the Old World by seas and oceans. The hominids were tropical animals that arose from ape ancestors in this semi-isolation. Why they should have ever left Africa is as important a question for human evolution as any phylogenetic or anatomical conundrum that has traditionally occupied paleoanthropologists.

The impelling force for *Homo erectus* populations to extend their range outside Africa was the same ecological force that caused them to expand the sizes of their home ranges. As climate became more arid, food resources shrank. As food resources shrank, hominids were forced to move farther and farther afield in search of protein and plant foods. If one covers a bigger territory and keeps on the move, chances are that something to eat will turn up. We surmise that this happened to *Homo erectus* because of the increase in body size of the species, which correlates in modern mammals with increased home range size.

By chance, the hominid groups in northeastern Africa moved into southwestern Asia across the Isthmus of Suez by this process, but one can be sure there were no signs welcoming them to a new continent when they crossed over. The pioneering *Homo erectus* were certainly just concentrating on searching for food resources in an area that did not look or feel that different from the area from which they had come.

Ecologists would describe the diet of *Homo erectus* as "high-quality" food that was "widely dispersed." Hominids would have searched for high-calorie, high-protein "packages" of food, like animals or particular plant parts. The patches could have fruiting trees; watering holes where animals congregated; an area of moist ground where tubers grew; a pool that periodically dried up, stranding fish; or a cliff where migrating young antelopes might fall—any place that food could be found. This dietary adaptation has been called omnivorous, scavenging, hunting, hunter-gatherer foraging, and other combinations of these terms. It is very different from the dietary adaptation that typifies herbivores, which eat "low-quality" but abundant food resources like grass. Because antelopes, for example, have plenty of grass to eat, and it is of predictable abundance, many of them live in one relatively small area and their population density is high. Hominids are much like carnivores in respect to the amount of territory they cover in their quest for food. Like lions, leopards, cheetahs, hyenas, jackals, and hunting dogs, hominids had a species range that crossed from Africa to Eurasia, whereas species of African and Eurasian antelopes and other herbivores are much more diverse because of their tendency not to disperse so widely.

A major force pushing hominids out of Africa was simply the pressure to find greater food resources. This reason, or one similar to it, has sufficed for most paleoanthropologists who have wondered about the phenomenon. But there are several implications of this simple model that do not accord very well with other observations, necessitating some additions to be added to the theory of *Homo erectus* dispersal out of Africa.

First off, the outlet from Africa to Eurasia was, and still is, very small. Looked at in another way, there was a major water barrier, composed of the Mediterranean and Red Seas, between Africa and Eurasia. Passive dispersal of hominids (and other mammals such as carnivores) from Africa to Eurasia would have been a trickle over this tiny land bridge. There would have been some, but not much, gene exchange between populations in Africa and Eurasia. Thus, we might expect the *presence* of both carnivores and hominids on both sides of the Suez, but we would also expect a fairly significant degree of differentiation of African and Eurasian populations as evolutionary lineages took their own courses in the two areas. However, we see a close correspondence in species. African and Asian lions, leopards, hyenas, and dogs are the same species, and early African *Homo erectus* looks very similar to the mainly later *Homo erectus* in Asia. There must have been a greater pressure to the pump that pushed out these species through the narrow Suez nozzle into Eurasia.

The force that pushed wide-ranging savanna-adapted species out of Africa at the beginning of the Pleistocene was climatic change. The strips of savanna, sahel, and desert that cover modern Africa in bands that fan out northward from the central forests expanded when global climate became drier. We can think of this as a physical pump. The savanna-sahel band formed the piston of the pump that moved northward from the tropics as climatic aridity progressed, compressing species' ranges before it. Woodland and forest species in the northern half of the continent were driven to extinction, caught between a rock (the advancing savanna) and a hard place (the Mediterranean and Red Sea basins) as their habitats disappeared. For species that had become adapted to savanna, climatic aridity at the end of the Pliocene actually enlarged their potential ranges in Africa, but even these species were pushed by the more arid advancing sahel at the beginning of the Pleistocene. It was this paleoclimatic pump that added the pressure needed to account for large numbers of African savanna animals, including *Homo erectus,* dispersing across the narrow Isthmus of Suez. Paleontologist Eitan Tchernov of the Hebrew University of Jerusalem records this mass influx of African mammals into southwestern Asia at the beginning of the Pleistocene. Even today, Israel's Negev Desert recalls the flora, fauna, and climate of northern Africa.

An important observation is that australopithecines never seem to have made it across the Suez to Eurasia—not the earliest australopithecines, who were within spitting distance of the Arabian peninsula, or even the later, robust australopithecines who were under the same aridifying climatic conditions as *Homo erectus.* Passive dispersal clearly did not take early australopithecines to Asia, probably because the climatic corridor was too open and too arid for them to cross. The paleoclimatic pump apparently did not push into Eurasia the robust australopithecines, who so far as we know were occupying the same habitats as *Homo erectus.* Why not?

Homo erectus must have had some advantage that allowed the species to adapt to more arid conditions than the robust australopithecines. From a strict anatomical perspective, the robust australopithecines were more adequately endowed in this regard—with their huge crushing molar and premolar teeth. It is a straightforward deduction then that whatever ecological advantage *Homo erectus* had was resident outside its body or perhaps has remained hidden from anatomists in its more capacious cranial cavity. It is possible that this advantage was a superior type of stone tool, a new type of wooden or bone tool, roofed shelters, clothing of some sort, new cooperative food-getting techniques, or other social adaptations that australopithecines did not have. The evidence does not yet extend this far, but my guess is that fire was the critical cultural adaptation that gave *Homo erectus* the ecological means by which to disperse to Eurasia when the australopithecines could not. Having first evolved in the tropical savanna as a seasonal mechanism for food-getting, fire use was a "preadaptation" for life in the even more seasonal continental landmass of Eurasia. It eventually allowed an adaptation to cold conditions unthinkable for any other primate.

DEATH OF THE NUTCRACKER PEOPLE

If increased intelligence and the use of fire allowed *Homo erectus* to evade the pincer movement of climatic change at the onset of the Pleistocene, its bull-necked hominid relatives were not so fortunate. The robust australopithecines went extinct at about this time. The latest fossil records of the robusts are about 1.2 million years ago in eastern and southern Africa. This date is about the top of the Omo fossil sequence, the longest record in Africa. Sites immediately after this point in time are lacking. Because of this gap in the fossil record, it cannot be said with certainty that these massive-jawed hominids disappeared until about three-quarters of a million years ago. This is the date of the large *Homo erectus* site

of Ternifine in Algeria, which has yielded no fossil relics of our robust cousins. Some time before this, robust australopithecines had
gone extinct.

Trying to figure out why a fossil species may have become extinct is dicey business, unless we have a smoking gun. For the numerous extinct species like the dodo bird, the Tasmanian wolf, the
passenger pigeon, and the quagga, we do. These species were
killed off by human hunting or ecological depredation within historical times. Could the robust australopithecines have been the
first of humanity's ecological victims, or is this a case of simply extending an explanation from the historical present to the past because we are bereft of other ideas?

Several reasons make it likely that on general principles *Homo
erectus* would have been an important ecological competitor of
the robust australopithecines, mainly because they were closely related species and because they occupied virtually the same habitats. Their ecological niches then would have been very closely
spaced. If anything had happened to decrease the size of the robust australopithecine niche, or to increase the size of the *Homo
erectus* niche, the entire niche space of the robust australopithecines would have been overlapped by *Homo erectus.* The theory of competitive exclusion holds that when there is total niche
overlap by two species, one of them will eventually go extinct.

The use of fire by *Homo erectus,* which is well documented for
the species at such late sites as Zhoukoudian, may well have been
the factor that created total niche overlap between *Homo erectus*
and robust australopithecines. What the robusts had eaten by
brute force before, *Homo erectus* was now able to eat with the use
of fire. We do not know what these foods were. They could have
been certain plants, seeds, or animals. Whatever they were, the robusts needed them, and without them they starved.

It is tempting to speculate that *Homo erectus* might have even
preyed upon robust australopithecines, but there is no evidence
for this. The very similarity in adaptation that brought robust australopithecines into competition with *Homo erectus* may have

made them tough adversaries in any direct competition. Even to-day, *Homo sapiens* hunting for food steer clear of apes, whose strength and intelligence make them formidable prey. I recollect running across a group of hunters in a remote section of eastern Zaire inhabited by chimps. They had just run down a bush pig and I asked one of them if they ever shot chimps. He shook his head vigorously and pointed to his dogs. The *soko mutu* will tear the dogs to pieces with their hands, he said. The evolutionary dis-tance between *Homo erectus* and robust australopithccines was significantly less than that between humans and chimps, and one can suspect that robusts were big enough and smart enough to fight back. It is most likely that the robust australopithecines went extinct as a secondary effect of ecological competition with *Homo erectus,* occasioned in large part by the significant climatic changes that accompanied their disappearance.

Homo erectus stands at the threshold of humanity. The species represents our last contact with our australopithecine past as well as our new conquest of continental Eurasia. But that very near-humanity could well make us look pejoratively upon *Homo erectus* if we were to conjure one up out of the mists of the past. We can accept chimps (and by extension australopithecines) "hunting" animals—including members of their own species—by tearing them apart with their hands because this is "animal behavior." Somehow *Homo erectus* should have known better because in a vague way it was "human." But this head-bashing, cannibalistic, cave-dwelling, smoky-smelling, and slavishly imitative hominid was a far cry from our concept of "human." It was the remaining cataclysm of Pleistocene climate change that was to shape this clay into the large-brained species that we know as *Homo sapiens.*

8

Neandertals and Mitochondrial Eve
Dance a Pas de Deux
to the Rhythm of Climate Change

Long before Fred Flintstone, B.C., and even Alley Oop, the "cave man" had stirred popular interest. One of the first motion pictures, by D. W. Griffith, dramatized the primordial competition between aquiline-featured modern human beings, our ancestors, and the brutish and coarse Neandertals, replete with clubs, bad posture, and female mates dragged by the hair. The movie represented a mixture of myth and science that had grown up over the past half century around the strange fossil discoveries of Neandertals in Europe. Its plot, however, was a popular rendition of a major paleoanthropological discovery of only a few years before.

Otto Hauser, a professional collector and excavator from Germany who worked around the turn of the last century, discovered in 1908 a fossil skeleton of an adolescent Neandertal at the cave of

Le Moustier in France. Associated with the skeleton were Mousterian stone tools—relatively large flaked tools used for spear
points and scrapers. Directly above the stratum of the excavation
in which the Neandertal was found—that is, immediately after the
Neandertals had lived in the cave—there were abundant Aurigna-
cian tools. Hauser, as well as all the archaeologists of the day,
knew that these smaller and more refined tools were only associated with anatomically modern people, the so-called Cro-
Magnons. The conclusions were obvious to Hauser. The primitive
Neandertals were displaced, perhaps forcibly, by the more
anatomically and culturally advanced Cro-Magnon people.

Paleoanthropologists have been working with this information
and similar discoveries ever since. Still today there are specialists
who think that Hauser was essentially correct—the Neandertals
were entirely and completely wiped out and superseded by
anatomically modern humans. And then there are those who believe that Neandertals were ancestral to, and evolved into, modern humans in Europe and the Middle East. Both scenarios have
been around a long time, both have zealous supporters, and both
have inherited a number of old ideas from the past century. Traditionally these two viewpoints have been considered diametri-
cally opposed to each another, but from a standpoint of
evolutionary biology, paleoecology, and our new multidiscipli
nary focus, they need not be, at least entirely. This chapter will
put these ideas in a different light. The Neandertals did go extinct as a population of *Homo sapiens* when Cro-Magnons took
over Europe, but it is also likely that many Neandertal genes persisted in the gene pool.

Many of the old ideas about human evolution were extremely
arrogant from a standpoint of natural history. Theoreticians imputed to human precursors an ability to control evolution.
Friedrich Engels, for example, in his "labor theory" of human origins, thought that our ancestors had created themselves by working hard and cooperating in social communes. Sigmund Freud
believed that human behavior had its roots in a volitional act of

patricide by oedipally motivated brothers, and that in some way this psychological mindset had been passed on to later generations. Many early paleoanthropologists believed that human ancestors had basically "thought" themselves to modernity—the brain had in some way been the driving force behind evolution. In all of these formulations human evolution was regarded as the primary result of human action, sort of like history. These ideas are arrogant because they all assume that human beings themselves provide the spark, the impetus, for change. In this context, the Neandertal–Modern Human question comes down to a simple knock-down drag-out struggle between two contesting sides—sort of like the Philistines and the Israelites, the settlers and the Indians, or the Hutus and the Tutsis.

Although it is more than a little scary to contemplate that conditions and events outside our control formed and shaped our evolutionary history, this is exactly what a century of evolutionary biology teaches us. It is the theme of this book. The fanciful narratives of human origins have not looked at human beings as part of nature and subject to its laws. On the other hand, it was Charles Darwin and Gregor Mendel, scientists respectful of nature's laws, who humbly studied barnacles and pea plants, who put forward the theories central to understanding human origins. Other evolutionists, such as Ernst Haeckel, imputed a unique sort of vitalistic "will" to humans that they denied to all other species. But we have no evidence to indicate that human evolution was fundamentally different from that of any other species.

Like other species, humans arose in response to changes in the world around them. Thus, Neandertals and anatomically modern humans evolved and adapted to environmental changes just like the other Pleistocene animals and plants around them. Their populations expanded and contracted as conditions fluctuated during this climatically variable time. An ecological and nonanthropocentric perspective provides us with our viewpoint for interpreting the dramatic events of late Pleistocene human evolution.

The Origin of Modern *Homo sapiens*

As with so much of traditionally Eurocentric cultural and historical interpretation, the scientific story of the origins of modern humans, which first began to be discovered in Europe, has had to be
broadened to a global perspective. Fossilized human remains that
demonstrate substantial antiquity for anatomically modern people
were turned up in various places in Europe well before the publi-
cation of Darwin's *Origin of Species* in 1859. Their significance
could only be appreciated much later and in the overall context of
human evolution, since by themselves they did not document evolutionary change—only that modern people had lived a long time
ago. No one knew how long ago. It is now clear that European
anatomically modern *Homo sapiens* constituted only the edge of a
much larger and more ancient distribution of anatomically modern people that had spread into Europe, displacing or evolving
from the Neandertals. Nevertheless, the European fossil record of
late Pleistocene hominids is still the most complete and still exerts
a strong influence on our hypotheses of *Homo sapiens* evolution.

Currently there are two competing schools of thought concerning the origin and subsequent evolution of *Homo sapiens*. The first
of these is the "multiregionalist" school. Paleoanthropologists of
this persuasion believe that in any one area of the world there is
primary continuity from earlier populations to later ones. Thus,
early *Homo sapiens* fossils found in Europe, Asia, or Africa have a
probable ancestral relationship to later *Homo sapiens* fossils found
in those same areas. Multiregionalists believe that the genes that
are shared among human populations have moved by a process of
diffusion, a result of many matings between contiguous populations over time. This school of thought tends to have as its adherents those paleoanthropologists who have tilted toward the theory
that Neandertals, the "brutish" but big-brained *Homo sapiens*
that lived in Europe and the Middle East, were ancestors of
anatomically modern *Homo sapiens,* at least in Europe.

The other school of thought is known as the "replacement"

school, sometimes also called the "Out-of-Africa Theory." Paleo-anthropologists who hold this point of view believe that in many or most cases involving *Homo sapiens* evolution, major advances occurred in one region and spread to other regions by large population movements. Because Africa preserves the oldest recovered fossils of anatomically modern *Homo sapiens,* Africa has been the preferred region of origin for this migrationist school. These theorists hold that preexisting populations in regions outside Africa were replaced and therefore were not ancestral to descendant populations in the same area. They tend to believe that genes were carried by migrating people who moved into a region, and not so much by genetic admixture over time.

Were the Neandertals a separate species, unable to interbreed with Cro-Magnons to produce viable and reproductively capable offspring, or were they a separate population within the same species, different in appearance and behavior but capable of passing their genes to their anatomically modern cousins? The migrationists tend to adopt the former position and refer to Neandertals as *Homo neanderthalensis.* Multiregionalists gravitate toward the latter proposition and refer to Neandertals as *Homo sapiens neanderthalensis.* Several lines of evidence support the hypothesis that Neandertals were a distinct population of *Homo sapiens* but not a separate species. First, there is evidence that clinal variation took place (a *cline* is a pattern of gradually changing traits along a geographic gradient). Neandertals from eastern Europe look less "classic" and more modern than do Neandertals from western Europe. This implies that gene flow with anatomically modern populations to the east was occurring. Second, the categories of anatomical difference noted by generations of anatomists between Neandertals and anatomically modern *Homo sapiens* are significant and admittedly greater than we see in "races" of modern humans, but they are still not of the magnitude that generally separate different species of animals. Finally, the degree and duration of separation of geographic populations necessary to split one formerly continuous population is unlikely to have occurred in the case of Neandertal evolution. As we will see in this chapter, mixing and churning of faunas and floras in Europe throughout the Pleistocene probably never allowed a total genetic cutoff of populations of large, mobile animals like *Homo sapiens.* It is not likely therefore that Neandertals were a separate species and they will thus be considered here as *Homo sapiens neanderthalensis.*

The debate between the multiregionalists and the Out-of-Africa theorists has been cast in the traditional paleoanthropological model of contesting theories of descent. The fossil evidence, both pro and con, has been debated at length, but the real questions here are ecological. If we could understand the timing and placement of human-environmental interactions, we could determine whether one or the other scenario is more likely correct. This will be our starting point in unraveling the evolutionary history of modern *Homo sapiens.*

HIPPOS ON THE THAMES AND DISHARMONIOUS FAUNAS

Paleontologists in western Europe recognized early on that something major had occurred to scramble up climate and mix faunas during the Pleistocene. Tropical animals like hippopotami, elephants, and ostriches were found far to the north of any place that they had ever been reported or known to have existed in historical times. With little temporal control of their discoveries, the paleontologists could conclude simply that tropical climates had existed in northern and western Europe during the Pleistocene. A fossil hippopotamus discovered in Pleistocene river gravels in England showed not only that this tropical large mammal had found an acceptable habitat here during the Pleistocene but also that this typically African species had dispersed the more than 5,000 kilometers from its continent of origin. At the same time, species from the far north, such as reindeer, were common at a number of Pleistocene sites in western Europe, and their bones and antlers had been used by humans to make tools. From the faunal evidence, the Pleistocene seemed to be a mishmash—an epoch of confusing ecological contrasts.

In North America, paleontologists discovered the same phenomenon in their Pleistocene sites—a mixture of tropical faunal elements with cold-region species. At first these strange mixtures of fauna, which concatenated moose and elk fossils with those of armadillos and Galápagos-like tortoises, were chalked up to sloppy stratigraphy. It was assumed that the animals from different climatic regimes must represent different, albeit closely spaced, periods of deposition. Only recently, after controlled excavations with well-established geological stratigraphy, have paleontologists realized that all these disparate animals really did live together at the same time and in the same place. Because of the incongruous ecological adaptations of the animal species contained within them, these Pleistocene faunas were termed "disharmonious." Understanding how they came about is of major importance in understanding modern human origins and evolution.

From our consideration of the paleoclimatic pump that pushed *Homo erectus* and some large mammals out of Africa at the beginning of the Pleistocene, it is clear that a mechanism was already in place to periodically push out from Africa both hominids and other animals. But it is also true that African faunas do not appear in Europe until the later parts of the Pleistocene. The earliest Pleistocene emigrants from Africa appear in Israel, and only later in sites to the north—in central and western Europe. Similarly, only in the later sites in Pleistocene Europe do we see the emigrants from the far north.

The seeming paradox of tropical species mixing with species from the far north makes sense if we look at the pattern of Pleistocene climatic change recorded in the deep-sea isotope curve (Figure 10). As the Pleistocene advanced, two major climatic trends can be seen. The first one is well appreciated as the generally lowered mean temperatures that accompanied the advance of the glaciers and the "ice ages." The second aspect of Pleistocene climatic change, of even more significance but frequently not appreciated, is that the amplitude of change became increasingly greater. The colds became colder and warms became warmer, on the average every 100,000 years between major peaks. The shifts of vegetational belts along lines of latitude became extreme, carrying their faunas with them. Near-tropical belts were shifted far to the north during the warm phases, and near-arctic belts were shifted far to the south during cold phases. During the transitional periods, between extremely warm and extremely cold conditions, there would have been mixed faunas—for example, warm-adapted species left over from the preceding warm period and cold-adapted species migrating in with the advancing cold period. During the cold phases, both warm-adapted and cold-adapted faunas would have been pressed between the southward-moving glacial front and the northward-moving arid front, thus placing these species in a climatic vice and pitting them in a competitive ecological relationship with each other. Human populations, originally part of the tropical faunas, were caught in this same Pleistocene vice and their evolution was profoundly affected.

MITOCHONDRIAL EVE AND MODERN HUMAN MIGRATIONS OUT OF AFRICA

The record of past human history that can be read from the ground—the fossil and archaeological record—is limited. This is true even for the Pleistocene, which preserves the most complete record because it is the most recent, its sediments are the most common, and its deposits have undergone the least geological disturbance. Paleoanthropologists have not been able to resolve the debate over the geographical and temporal origins of *Homo sapiens* from this record alone. An entirely different line of evidence has come into the equation. Molecular biology has over the last fifteen years contributed a major corpus of data on the origins and evolution of modern humans. These data and the conclusions derived from them have not always been appreciated or even understood

Figure 10

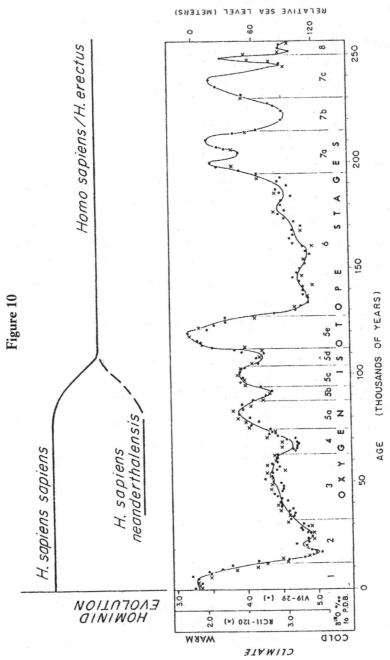

The oxygen isotope curve for the latter part of the Pleistocene. Neandertals are recorded from Europe during and after Stage 6, implying that they may have been isolated by environmental and paleogeographic factors related to this extended cold period (from Boaz et al., 1982).

by the scientists trained in studying fossil bones, teeth, and archaeological artifacts. But fortunately, they have not been able to ignore them.

Biochemist Allan Wilson, who started the furor with Vincent Sarich over the ape–human split in the 1960s using primate immune responses to various proteins, was responsible in large part for generating the molecular controversy over modern human origins. Wilson knew that because DNA codes for the amino acids that make up proteins, the proteins that he had analyzed in assessing the ape–hominid split reflected changes in DNA—the greater the difference between a human albumin protein and the albumin of any other primate, the greater the DNA and evolutionary distances between them. The system worked well for grand questions of primate evolution, when long spans of time were being investigated. But for questions of human evolution, within the last few million years, there was not enough accumulated change in the DNA contained within the nuclei of the cells to accurately measure change. It was a bit like using a sundial to time the 50-yard dash. Wilson needed a faster-changing molecule to measure human evolutionary change.

Wilson and Rebecca Cann, one of his graduate students at Berkeley, chose a type of rapidly evolving DNA molecule found inside most of the cells of the body, but outside the cell's nucleus. This extranuclear DNA is found within an organelle of the cell, the mitochondrion, and from comparisons with different species whose times of evolutionary divergence were known, its rate of change was estimated to be ten times that of the nuclear DNA. If Wilson and Cann could analyze the sequences of mitochondrial DNA (mtDNA) from many different human populations around the world, they could calculate when they all may have shared a common ancestor. This would date the origin of *Homo sapiens* without ever having to stick a spade in the dirt or visit a single museum collection of hominid fossils.

For Wilson and Cann, getting the mtDNA data was a lot easier, as it turned out, than interpreting what it all meant once they had it. The sequences of chemical bases that make up DNA are very long, and assessing changes from one to another is not an easy task. Sometimes there is clearly one changed position between populations 1 and 2 (Figure 11). This difference indicates that a change took place in one or the other of the evolutionary lineages of the populations, but which one? The mtDNA sequences of the other populations around the world had to be looked at to figure this out. If almost all of the other populations had the same base as that of population 1, then one could safely conclude that this was the primitive state of the DNA and that population 2's change was a more recent mutation. Since there were many such differences, Wilson and Cann used a computer program to analyze all the various sequence differences in their mtDNA study.

Using more or less the same molecular clock theory that Wilson and Sarich had used fifteen years before, Wilson and Cann concluded from their mitochondrial DNA analyses that all their samples could be traced back to one point in time. Their computer program drew them an evolutionary tree from all its one-by-one comparisons of mtDNA sequences and it converged on one time—about 200,000* years ago. Their data showed all living people had shared the same ancestor. From the geographical location of the mtDNA samples that showed the highest retention of the original sequence, Wilson and Cann concluded that the site of modern human origins was in Africa. Pressed by scientific journalists that needed a hook for their stories, Wilson and Cann noted that mtDNA was passed on only in the female line, and that, yes, ultimately the mtDNA configuration would have arisen in one individual female, one who lived in Africa some 200,000 years ago. Thus, "Mitochondrial Eve" was born.

Mitochondrial Eve is inordinately important in our under-

---

*The original range of dates proposed by Wilson and Cann was 140,000 to 290,000 years ago. Because of the wide range of these dates, Cann more recently (1993) has used "about 200,000 years ago."

Figure 11

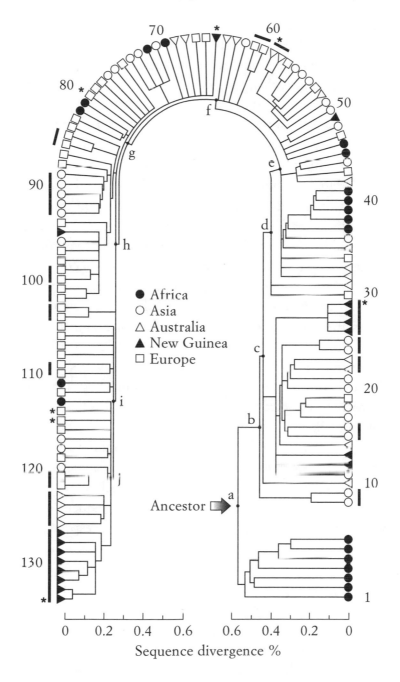

Mitochondrial DNA phylogeny for the human species (based on work by R. Cann and A. Wilson; from Boaz and Almquist, 1997).

standing of modern human evolutionary origins—not as some semimythological African fertility priestess who promulgated the entire human species, but as a metaphor for a small population of closely related individuals who shared a particular configuration of mitochondrial genes that was subsequently passed on to the entire surviving human population. The fact that this configuration of genes is traceable back to an African source 200,000 years ago is important but it is not by any means the end of the story. Other genes and DNA configurations will likely be found that have their earliest origins in other parts of the world. It is important to remember that gene lineages and organism lineages do not always correspond. But we would expect that *most* of the gene complexes of modern humans will trace back to Africa if this continent did indeed provide most of the migrational impetus behind human dispersal over the rest of the Old World. Thus, Mitochondrial Eve supports the paleontological and anatomical arguments for the "Out-of-Africa" replacement theory but it does not provide incontrovertible proof.

An ecological scenario offers a further framework for assessing the mitochondrial evidence for recent human evolution. The assumption of Wilson and Cann that there was one, and only one, "splitting" event that led to the formation of modern *Homo sapiens* is likely an oversimplification. While this total cutoff from the genetic spigot may work for the splitting mechanics of species, when it can be assumed that there was no subsequent gene flow between populations after they diverged on their respective evolutionary lineages, this assumption cannot work for populations of the same species. By definition, populations within the same species can share genes, most frequently by direct gene exchange through matings at the edges of their population distributions. Some of this sort of gene exchange could have muddled the dates of the origins of modern humans that Wilson and Cann calculated, but it would not change their overall conclusion that Mitochondrial Eve was African and that "she" lived some time during the late Pleistocene.

The recurring waves of migrational movements of faunas and humans in the late Pleistocene correspond to low points on the oxygen isotopc curve (Figure 10). These low points occurred at 260,000 years ago (Stage 8), 220,000 years ago (Stage 7), 130,000–170,000 years ago (Stage 6), 70,000 years ago (Stage 4), and 20,000–30,000 years ago (Stage 2). The timing, length, and magnitude of Stage 6 make it the most likely candidate for having been the most important cold/arid climatic phase in pushing the "Mitochondrial Eve" population out of Africa, but there would have been other, probably smaller, dispersals that could have clouded the timing of the genetic picture.

Some critics of the Mitochondrial Eve interpretation have argued that population dispersal could have just as well occurred the other way around—that populations from Eurasia migrated into Africa, thus accounting for the shared genes, and that Wilson and Cann got it all wrong. Despite the facts that the modern genetic diversity of *Homo sapiens* in Africa dwarfs that of the rest of the world, and that the earliest fossil records of anatomically modern *Homo sapiens* occur in Africa, the paleoclimatic data do not indicate a clear mechanism for this type of population movement. Increasing aridity in Africa and southern Asia would have forced populations to the north as climate deteriorated, but when equable conditions returned with cooler and wetter climate, there would have been no reciprocal force pushing animal and human populations back across the Suez into Africa.

Instead, populations already present in an area would tend to disperse more widely as acceptable habitats expanded. There is some evidence for this type of population expansion later in the Pleistocene. The "paleoclimatic pump" then was most likely a one-way pump pushing species out of Africa but not reversing itself when climate returned to more equable conditions. The paleoecological evidence here supports the Mitochondrial Eve model, but it would indicate a somewhat later date for the main dispersal of the Mitochondrial Eve population out of Africa, probably during Isotopic Stage 6, about 130,000 to 175,000 years ago. These dates are also in better agreement with the absolute dates on African sites such as Omo Kibish and Klasies River, which have yielded remains of anatomically modern humans, both at about 100,000 years ago.

If the fossil, biomolecular, and paleoecological data correspond well enough for us to accept the Mitochondrial Eve model during Stage 6, what of the other late Pleistocene climatic episodes of nearly equal intensity both before and afterward? Would we not expect similar ecological conditions at these times? Going back to about 1 million years ago, when the 100,000-year cycle became predominant, there were another seven or eight major cycles of global cooling and aridity. We do not yet have accurate data on most of these time slices, yet it seems that at about 500,000 years ago we have the first appearance of archaic *Homo sapiens.* These mechanisms of mixing of faunas undoubtedly had a large effect on homogenizing the Pleistocene human populations throughout most of the Old World. Subsequent to the Mitochondrial Eve dispersal, the climatic changes characteristic of Stages 5, 4, and 2 played a formative role on modern human evolution, including the demise of our close cousins, the Neandertals. We turn to this subject next.

RESOLVING THE NEANDERTAL PROBLEM

When I was a graduate student preparing for my Ph.D. oral exams, I remember reviewing the answer to one of the standard questions: "What is your position on the 'Neandertal Problem'?" The part of the question that gave me the most trouble was the "the" part—what, I wondered, about the Neandertals was *not* a problem? What functional interpretation could one ascribe to their morphology—the heavy browridges, the long but quite capacious skull, the protruding bony nose, the short and squat body build, the uniquely thin pubic bone in their pelvis? There was not a consensus then and there are still no conclusive answers.

How Neandertals *used* their anatomy was also a problem. The front teeth in several specimens, for example, showed an unusual type of rounded wear. Were Neandertals chewing their mukluks like Eskimos, and if so, why with their *front* teeth, or were they using their front teeth as a vice to hold implements they were fashioning? What sort of life did the Neandertals lead? They were so close to modern humans in brain size, they buried their dead, and they apparently believed in some sort of religion, yet they left no record of pictorial art or carving that typified their Cro-Magnon successors. How had the Neandertals come to be a separate lineage from mainstream *Homo sapiens?* What had happened to them between 32,000 and 40,000 years ago that led to their disappearance? Were there Neandertal genes in surviving European populations or did *Homo sapiens neanderthalensis* go extinct without issue?

One of the reasons that the Neandertals have been such a problem for traditional paleoanthropology is that their origins and extinction are basically paleoecological problems—problems that cannot be solved by primary recourse to anatomy, the strong suit for most paleoanthropologists. To be sure, Pleistocene geologists as far back as Charles Lyell, a senior contemporary of Darwin's, had investigated the age and depositional contexts of the Neandertals. But until three scientific developments had occurred, resolution of the contextual issue surrounding the Neandertals could not be reached. Accurate absolute dating methods, reliable methods of assessing past climates, and a population approach in evolutionary biology all had to be developed before any real progress could be made.

A Time Line for Neandertals

Paleoanthropology is in large part a reconstructive science—it puts together the pieces to solve a puzzle. In this aspect it is not unlike a criminal investigation. In both, the sequence of events is critically important to establish before an accurate overall picture can be reconstructed. We of course cannot put a Neandertal on the witness stand and ask him where he was on the night of April 9, exactly 50,000 years ago. Instead we must interrogate the earth.

Radiocarbon dating has been the primary technique that paleoanthropologists have used for the last forty years to date the most recent phases of human evolution. Half of the carbon 14 present in a organism that has just died decays in 5,700 years. After each 5,700 years (the half-life of carbon 14), the amount to measure is reduced by half. Eventually a point is reached where there is so little carbon 14 left that it is virtually impossible to detect. This point, usually at about an age of 50,000 years, is the limit of standard carbon 14 dating.

Unfortunately for piecing together the Neandertal story, the lower age limit for conventional radiocarbon dating falls right in the middle of the heyday of *Homo sapiens neanderthalensis*. It has been impossible to tell when the first Neandertals came onto the scene or to plot their population movements and evolutionary change through time. Only the date of their demise has been accurately determined—32,000 years ago in western Europe and 40,000 years ago in the Middle East. These ages are radiocarbon dates that mark the last of the Neandertal fossils before the appearance of anatomically modern humans.

The development of new dating techniques has extended our accurate knowledge of the Neandertal period of time. Accelerator radiocarbon dating employs a cyclotron or particle accelerator to count individual atoms of decaying carbon 14. It is very accurate and can be run on a much smaller sample (which is destroyed in the process). Larger samples can be run to determine dates older than the standard limits of radiocarbon dating, but because the technique causes fossil destruction, most anthropologists have opted for other newly developed methods.

The most important of these new techniques is electron spin resonance (ESR). This method, like radiocarbon dating, measures decaying elements (primarily uranium, thorium, and potassium) and calculates an age on the basis of the rate of decay. Also like radiocarbon dating, ESR dates the fossil itself (versus the enclosing sediment). Usually the material dated is tooth enamel, which acts as a natural high-density trap for the electrons spun off from the decaying uranium or other element. By "resonating" the trapped electrons between a strong magnetic field and a beam of microwaves in a machine called an ESR spectrometer, the radiation trapped in the sample is released and measured. Ideally, the greater the amount of energy released, the older the sample.

Unlike radiocarbon dating, where we can assume with some confidence what initial amounts of carbon 14 were, the initial concentrations of the radioactive elements used in ESR dating vary from site to site and have to be measured at each place. A number of corrections and precautions have to be observed in ESR dating, but since the mid-1970s, methods have been greatly improved. Several laboratories now obtain consistent and reproducible dating results. ESR dating has an effective range from about 10,000 years ago back to perhaps over a million years ago, although its main use so far has been in that period of time overlapping and predating the effective time range of radiocarbon dating—the time of Neandertal and modern human origins.

Rainer Grün of Cambridge University undertook ESR dating of the classic Neandertal sites of La Chapelle aux-Saints in France and Monte Circeo in Italy. He found that these populations lived between 50,000 and 60,000 years ago. Le Moustier, site of the pre–World War I discovery of the presumed replacement of Neandertals by anatomically modern people, was dated by ESR to significantly later. The Le Moustier Neandertal has a maximum age of 41,000 years ago, near the radiocarbon-determined age at other sites for extinction of the Neandertals. At the other end of the Neandertal age spectrum, the site of Ehringsdorf in Germany, which yielded fossils that some paleoanthropologists interpreted as early or pre-Neandertals, gave an ESR age of 230,000 years ago.

Bracketing most classic Neandertals in Europe between about 40,000 and 60,000 years ago was fully within the expected age ranges, so ESR dating was considered a confirmatory method of dates already established by radiocarbon dating and by relative dating of stages of mammal evolution (biostratigraphy). Even the early date for Ehringsdorf did not occasion much skepticism among anthropologists.

Electron spin resonance dating made big news in 1988 when an anatomically modern human fossil in Israel was dated to 100,000 years ago by Henry Schwarcz and colleagues in Canada. The site was Jebel Qafzeh, a cave near Nazareth that had yielded a well-preserved skull studied by French paleoanthropologist Bernard Vandermeersch. Vandermeersch's conclusions, published in 1981, had been that the specimen was modern in morphology. There was little doubt about his conclusion. But a date of 100,000 years ago for anatomically modern *Homo sapiens* in the Middle East was a whopping 60,000 years earlier than previously accepted. The fossil fauna supported the date, and so did the related dating method of thermoluminescence carried out on several burned flints found at the site. Subsequent ESR dating at the Skhul cave site, another classic site with anatomically modern humans in Israel, gave dates between 81,000 and 100,000 years ago, in the same age range as Qafzeh.

Well-known Neandertals from the Israeli sites of Tabun and Kebbara turned out to be later than Skhul and Qafzeh, at about 50,000 to 60,000 years old. And then, at about 40,000 years ago, the Neandertals were gone, replaced by anatomically modern populations discovered at the upper layers at Skhul and the site of Amud, along the Sea of Galilee.

This convoluted history of anatomically modern humans, then Neandertals, and then moderns again did not make sense in any straightforward multiregionalist scenario. It was very unlikely that Neandertals living back to 200,000 years ago evolved into anatomically modern humans at 100,000 years ago, then back into Neandertals at about 40,000 years ago. Israeli paleoanthropologists Ofer Bar-Yosef and Yoel Rak interpreted the dates to mean that Neandertal and modern populations migrated into and out of the

Middle East at these times. This interpretation does indeed seem most consistent with the new dates, but what ecological forces would have caused these populations of hominids to migrate in the first place?

RECONSTRUCTING THE HOMELAND
OF THE NEANDERTALS

The record of fossil pollen that scientists have discovered and pieced together over Europe is a detailed record of vegetational and inferred climatic change for the last 500,000 years or so. This record of wind-blown genetic debris of plants is preserved in the sediments laid down by lakes and bogs. It reveals the ancient plant life of a region.

The pollen record shows that during periods when the oxygen isotope curve dipped to colder temperatures, steppe and tundra conditions spread throughout Europe. Forests retreated up mountain slopes to montane refuges. Temperatures were significantly colder. In the intervening periods more temperate conditions returned. Today's climate in Europe is characteristic of one of these intervening warm periods.

Animal fossils also show that the fluctuating climate exerted a powerful influence on the mammal communities in Europe. As extremes of cold and then temperate conditions swept into Europe in advancing latitudinal fronts, faunas from different climatic regimes were mixed. Only relatively brief climatic optima allowed these disharmonious faunas to persist for a few thousand years, and then severe cold returned. As the mean temperatures of each of the climatic rebounds became more extreme through the Pleistocene, the climatic maxima of more and more species' adaptive limits were reached. Eventually it became too cold during the cold periods for a number of temperate species to survive and it became too hot during the warm periods for the cold-adapted species to hang on.

Geological evidence indicates that availability of water was not a major environmental factor in Europe during the Pleistocene. Meltwater from advancing glacial fronts even during cold periods provided a significant amount of surface water. This was a major difference from the arid lands to the south during glacial maxima and may also explain why animal populations that could adapt to the lower temperatures migrated into Europe during these times.

POPULATION MOVEMENTS AND THE DEMISE OF NEANDERTALS

If we put the newly constructed time scale for Neandertal evolution together with the paleoclimatic evidence, we begin to see what environmental forces caused both the evolutionary divergence of these unique hominids and their eventual extinction.

The major environmental factor accounting for Neandertals was cold. Since hominids, as we have seen, are basically tropical primates, hot weather, so long as there is adequate water, does not present a major threat to survival. The cold of the European Pleistocene, on the other hand, kept the home range of the Neandertals separate from that of the more advanced hominids to the east and south, who did not share their adaptations to the frigid north. An ecological minimum, in this case temperature, can be an important limiting factor in the distribution of many animal species. The Neandertals' unique adaptation allowed them to survive the climatic oscillations in Europe that brought on extreme cold. Squat of build and compactly muscled, they were the Eskimos of the Pleistocene, albeit with a more primitive and more physical mode of adaptation than their modern *Homo sapiens* counterparts. They used fire effectively and they took advantage of natural shelters such as caves to protect themselves from the elements. There are many nuances of adaptation that we can only guess at, but the 150,000 years or so that Neandertals occupied

Europe argues that their adaptation to the fluctuatingly frigid conditions of the ice age was successful. The very inhospitality of the environment during cold phases would have been an effective deterrent for more temperate-adapted human populations to move into Europe. But why, when conditions ameliorated, did not the larger human populations to the east and south move into Europe?

During periods of climatic equability in the European Pleistocene, when more temperate-adapted hominids *could* have moved into and occupied Europe, the Neandertals themselves may well have been an important factor in excluding other human populations. They were strongly muscled people who used brute force for many activities that *Homo sapiens sapiens* thought their way through. Neandertals, for example, have uniquely robust muscle attachments on their shoulder blades, lacking in anatomically modern populations, that indicate a very strong lateral rotation of the arm, as used in habitual and forceful wielding of a spear. Classic "in-fighters," Neandertals must have been a formidable force in hand-to-hand combat, be it with a cave bear in competition over a prime cave site, or with other human populations in competition for hunting/gathering territory. We have only a few bony evidences of intraspecies competition, but we can presume that the Neandertals successfully protected their borders against potentially migrating anatomically modern populations for the many thousands of years that they occupied Europe.

When conditions were warm and well watered, there was no real pressure for human populations in northern Africa and southern Eurasia to move into northern and western Europe, especially against the threat of direct competition with Neandertals. When their populations expanded, there were much wider horizons to the south and east than to the western cul-de-sac of Europe. Neandertals, on the other hand, just because they had no space to the north, west, or south in which to expand, tended to move out from their western European homeland from natural population growth pressure in times of plenty. This population movement is presumably what we have detected in the fossil record of Israel at 50,000 to 60,000 years ago when Neandertals displace anatomically modern *Homo sapiens.* Neandertals extended eastward out to what is modern-day Iraq at this time, but there is no evidence that they went farther south than Israel. Neandertals were almost certainly limited by their adaptation to cold and probably did not tolerate near-tropical conditions well.

At 40,000 years ago, Neandertals disappeared from the Middle East for the last time, just as climate was deteriorating into the last glacial maximum, an extremely cold phase that was not going to peak until some 18,000 years ago. The Middle East became hot and dry, a very inhospitable environment for Neandertals, and they retreated back to their central home range in Europe. Anatomically modern populations moved in to fill the void, and they brought adaptations for dealing with environments lacking abundant surface water. We do not know exactly what these adaptations were, but they were certainly primarily cultural and they had been born of necessity on the arid Pleistocene African savannas and the plains of Central Asia.

Had climatic conditions stayed at the equilibrium where they were about 40,000 years ago, the Neandertals may have survived and might still be with us today. But climate continued to deteriorate. Conditions outside Europe became drier, and as global temperatures dropped, the ability of human populations in the Middle East, North Africa, and southern Asia to survive in near-desert conditions also diminished. Europe, despite its cold, had water. It also offered large game animals that were becoming rarer and rarer in the surrounding arid areas. The conduits to Europe were also made more accessible. The expanding glaciers locked up more and more of the earth's water, lowering sea level tens of meters, so that land bridges around the Dardanelles, the Black Sea, and the Mediterranean were significantly widened. The human populations in the areas surrounding Europe then had ecological incentive and geographical opportunity to move into Europe in large numbers.

In recent years it has become fashionable for paleoanthropologists to suggest that anatomically modern humans displaced Neandertals because of their cultural superiority. This idea goes back to Hauser's day, but there is archaeological support for it. The tools of the modern humans that appear in Europe between 32,000 and 35,000 years ago are smaller, more refined, and made into many more regional varieties than the preceding Mousterian tools. There is a little mixture at the interface, however. Some "classic" late Neandertals are associated with the advanced tools, indicating that the correlation between physical type and cultural adaptation was not absolute. The most likely interpretation of these data is that, as modern human populations moved into Europe due to ecological pressure, they brought with them cultural attributes that they adapted to the extreme cold of the European Pleistocene. The endemic Neandertal populations were crowded ecologically and likely resisted incursions into their territories as they had done in the past. In this case, however, there was such a torrent of human population movement into Europe from the vast aridifying lands to the south and the east that the Neandertals were simply overwhelmed by the numbers.

As modern human populations successfully established themselves within Europe over a period of 5,000 to 8,000 years, there may well have been some gene exchange going along with cultural exchange. The successful cultural adaptation of the modern humans in Europe, however, allowed them to reproduce, dilute, and eventually swamp out much of the resident Neandertal gene pool. Thus, modern Europeans retain some Neandertal genes and they look the most like Neandertals of any extant human population (although Australian aboriginals do have on average larger browridges). They nevertheless trace a major kink in their family tree to the massive incursion of anatomically modern populations that occurred at the onset of the last glaciation.

THE EXTINCTION OF THE SOLO PEOPLE

The Neandertals represent the dynamics of late Pleistocene human evolution in the western extreme of their distribution as related to ecological change. As paleoanthropological work proceeds in other parts of the world, in Asia and Africa, we will hopefully be able to fill in the details of late Pleistocene human evolution in these regions with as much detail as for Europe. Africa is generally considered to be at center stage for most of the evolutionary advances in later human evolution. Asia, like Europe, seems to be more at the periphery of late Pleistocene evolution. Southeast Asia, for example, represents a region on the eastern extreme of the distribution of middle and late Pleistocene human populations. The record here suggests how what we have seen in Europe in terms of human population isolation and replacement may also be applicable to Asia.

Homo erectus was present in eastern and southeastern Asia in the early Pleistocene, and in fact Indonesia (Java) was where the species was first discovered. There is a major gap in human fossils and well-dated sites in Indonesia after *Homo erectus,* but a significant cache of late-Pleistocene fossil human crania, discovered in Java at Ngangdong on the Solo River by Ralph von Koenigswald just prior to World War II, does exist. These specimens are remarkably primitive for their apparent age and have always represented a conundrum to paleoanthropologists. Thought by some to be misdated *Homo erectus,* they have been generally shunted off to one side of the mainstream discussions of human evolution.

In view of the ecological perspective on Neandertal evolution just presented, it is likely that the Solo hominids really were off the mainstream of human evolution for quite some time. Isolated on the island of Java by rising sea levels after Isotope Stages 5 and 3, the human populations represented by the Solo fossils would have evolved in greater genetic isolation than even the Neandertals of Europe. They would have retained many of the features of their *Homo erectus* ancestors and thus would be among the most divergent of the late Pleistocene human populations known from the fossil record. When sea level dropped to less than 120 meters during the last glacial maximum (Stage 2), extensive continuity was reestablished with mainland Asian human populations. The Solo population suffered the same genetic fate as the Neandertals, but some of their genes perhaps survive in the indigenous populations of Australasia. We may thus plausibly explain both the robust crania of fossil human skulls found in the late Pleistocene of Australia as well as the large browridges of the extant aboriginals of Australia.

Expansion of human populations out of the Old World, sensu stricto, into Australia, the islands of the Pacific, and the Americas, represents an ecological tour de force. Humans did not accomplish this feat of dispersal over vast bodies of water and arctic conditions by means of morphological adaptations or by chance alone. Extracorporeal adaptations, in the form of culture, allowed these population movements. This uniquely human adaptation will be the subject of the next chapter.

9

Superorganic Culture Allows Modern Humans to Conquer the Ice Age and the New World

For people interested in cataclysms and punctuational events in human evolution, perhaps nothing is more arresting than the explosion of "culture" that occurred as climatic conditions became their most severe and inhospitable during the late Pleistocene glacial periods. This point was brought home to me during a seminar that I gave on hominid paleoecology in Boston. Stephen Jay Gould, the arch-punctuationalist, was in the audience. I expected a lot of flack because in general I do not think that human evolution is typified by long periods of almost no change and then sudden evolutionary "punctuational events." But it was not until near the end of the talk that Gould's hand went up. To my surprise, his question was not about any of the evolutionary biological "gradualist" conclusions that I had drawn. Instead, he wanted to hear more about the explosion of culture that occurred during the late Pleistocene. In retrospect, it made total sense. There is nothing in the paleontological record of the evolving human body that rivals the rapidity with which *Homo sapiens* began to evince advanced "out-of-body" culture—cave art, music, burial of the dead, clothing, personal ornamentation, diverse tools, and so on—between 20,000 and 25,000 years ago. If one is drawn to dramatic "hiccups" in the history of life on the planet, this certainly ranks near the top.

THE NATURE OF CULTURE

Before we can look at how and why human culture appeared in the latter part of the Pleistocene, we must first investigate what it is. Ask any one of the 5,000 or so practicing anthropologists and you will get about the same number of answers. Culture is the only intellectual glue that holds together the far-flung and tattered discipline of anthropology, and yet the concept is defined in many different ways, depending on one's discipline, training, school of thought, and particular theoretical bent.

In the past, cultural anthropologists and anthropological linguists, who study societies and languages cross-culturally, have claimed this central turf of anthropology as their own. But as archaeologists have dug deeper into the history of culture, and as biological anthropologists studying nonhuman primates have gained insights into the behavior of these near-relatives of humans, a much broader perspective on culture has begun to emerge.

A textbook definition of culture by two biological anthropologists* is "learned aspects of behavior passed on from one generation to the next in human societies." We all know that people must learn many, in fact most, things as they grow up to become

*Boaz, N. T., and A. J. Almquist, 1997, *Biological Anthropology, A Synthetic Approach to Human Evolution.* Upper Saddle River, NJ: Prentice Hall, p. 3.

effective members of their society. An important attribute of culture then is its ability to be assimilated by learning. Cultural attributes, such as a particular language, a specific way of determining who marries whom, how one throws a party, or even how one sneezes, are all learned. The fact that these behaviors are learned implies that they, or many aspects of them, can be changed, and changed as many times as necessary, if a society decides to alter what it teaches its members. This ability of culture to change rapidly the behavior of its members is a very powerful adaptive tool. It is the thesis of this chapter that the rapid adaptability of culture was the key element in human survival in the last phases of the ice age in northern Eurasia, as well as in the geographic dispersal of humans into diverse habitats throughout the world.

In addition to its rapid changeability, human culture also has the unique and paradoxical capacity to stay the same. A society's culture is not reinvented by the older generation every time its tenets are passed on. In fact, most of what is passed on in traditional societies is very close to what the older generation learned when they were young. How then does this aspect of culture jibe with its changeability? Perhaps Stephen Jay Gould's model of punctuated equilibrium applies much more aptly to cultural evolution than to biological evolution—long periods of stasis during which the status quo stays much the same from one generation to the next, punctuated by rare but rapid changes to a new order, a new way of doing things, a "revolution" in twentieth-century parlance. In whatever manner culture is passed on, however, it is faithfully reproduced in a functional way in the new generation. Culture is so important to human life in most environments that without it we would die.

The dual aspects of culture—its changeability and transmission across generations—are what separate it from the learned social behavior of the nonhuman primates. Human culture can *evolve*. Different groups of chimpanzees have been observed to have different learned behaviors, and some primatologists have suggested that these differences indicate incipient culture. But these primates do not depend for their survival on the learned behaviors that vary within the species. Rather, the behavior on which they depend is seen in all groups of the species, is invariant, and is thus closely tied to biological and genetic inheritance. There is intragroup transmission of the learned "protocultural" behaviors from one generation of chimps to another but there does not seem to be any change in the behaviors. Chimp learned behavior thus seems to be stuck in a perpetual "stasis" mode of evolutionary change, with little capacity for rapid response to changing conditions. This is perhaps a behavioral explanation for why chimps have remained in a tropical environment that is subject to little ecological change.

Human cultural behavior, in contrast, can apparently change on its own at a pace unaffected by any biologically imposed evolutionary rate. Cultural anthropologist Leslie White termed this attribute of culture "superorganic," implying that culture is beyond the laws of natural selection. This of course is impossible, since all behavior that affects reproduction, especially anything so pervasive as human culture, will affect biological evolution. But how could natural selection act to produce a mechanism of human behavior so tenuously connected to human biology?

THE EVOLUTION OF SUPERORGANIC CULTURE

Evolutionary biologists have undertaken experiments with fruit flies to determine how rapidly evolutionary change can take place. In the laboratory, spontaneous mutations occur with a frequency of about 1 in 10,000. Mutations supply the raw material for evolution, and if this figure is taken as an average mutation rate, we might imagine that it would take at least 100 generations in a population of 100 mating humans to generate the mutations for natural selection to *begin* to effect significant evolutionary change. One hundred generations is a period of some 2,000 years (assuming a generation time of 20 years), a pretty slow start-up speed for adaptation to a rapidly changing environment like ice-age Eurasia. What if environmental change was happening too quickly for human populations to meet the challenge, either by recourse to their range of physiological and behavioral adaptability in the short term, or by evolving biologically to meet the challenge—an option that implies tremendous mortality as natural selection worked over the long term? One answer is a mechanism of behavioral adaptation that has an inherently rapid evolutionary rate—culture.

Culture evolves rapidly because it is a positive feedback system. A new element created by the system does not slow it down (a negative feedback system) but instead makes it keep running or run even faster. The new element affects other parts of the culture, interconnecting with them and making the whole more complex. A hypothetical example illustrating this mechanism of environmental adaptation by culture might be the following.

> A tribe of ancient humans lived on the fringe of the glaciers in the Far North. A group of hunters tracks a herd of caribou that attempts to escape across a snowfield. A similar herd escaped from this same group of hunters the previous year because two of the hunters had gone temporarily blind from the intense glare of the sun on the snow. This year one of the group had carved eye masks that narrowed the light entering the eye to a single slit. The way the slits angled up looked somewhat like a fox. The hunters tied them on and were able to pursue the caribou across the snow until they managed to bag one. Their caribou fed the entire tribe during a time of the year when other foods were scarce, saving their group from near-starvation. Because the hunters had worn their new fox masks as a mark of triumph when they returned with their meat to the village, it was clear to the shaman that the spirit of the fox had guided them to the caribou. Henceforth, the shaman announced that he would consult the

spirit of the fox before each hunting party left the village. He planned a propitiation ceremony dedicated to the fox for the spring. The men in the original hunting party and their close male kin adopted the fox as their animal totem. This action effectively removed them from the wolf clan and meant that the impending marriage of one of the hunters' daughters to a man from the bear clan had to be postponed. Only after long discussion by the elders was it agreed upon that the fox and wolf were spiritually close, and that a member of the new fox clan would be permitted to marry a member of the bear clan, their traditional marriage partners.

This fictitious paleoethnographical vignette shows how a new element originating in a culture effects additional changes unrelated to it. These changes are not necessarily predictable other than to say that they feed back into the culture and become integrated. They make culture a constantly changing system that has a tendency toward increasing complexity. Furthermore, a new cultural element stimulates the addition of other new elements of culture. This very important attribute constitutes positive feedback, also termed *autocatalysis,* because in a sense culture creates itself.

If we accept the basic changeability of human culture as a means of rapidly adapting to a variable, dangerous, and potentially fatal environment, what is the value to a human population of all the apparently unrelated changes that go along with the primary cultural change in response to an environmental challenge? Why, in this instance, should wooden sunglasses lead to a religious ceremony dedicated to foxes six months later or have any effect whatsoever on the marriage plans of two adolescents? All this apparently wasted energy might in fact seem to be maladaptive. But it is not. Imbuing things and events with symbols and deep meanings is fundamentally human. It has a lot to do with how the brain works, as well as the mechanisms that human populations use to distribute food and resources, ascribe status, reaffirm family ties, and repay debts.

The Adaptive Value of Culture

From an external perspective, a potentially rapidly changing culture is a good adaptive platform on which to base a group's existence in an environment subject to rapid change. But from an internal perspective, what good is culture? How does it help an individual human being or a group adapt, survive, reproduce, and live the good life?

Culture consists of some basic ingredients—language, magic, religion, kinship systems, music, myths, accepted norms of behavior, dress codes, sexual mores, and rules for reciprocity. There are clear functional reasons, based on basic human biology, for a number of these universal cultural traits. For example, in the sexual more department, if there were not cultural rules that prohibited a man from having sex with a woman married to his neighbor (a man who had gone out hunting for the village), group cohesion would certainly break down when the hunter got back home and found out what was going on. Natural selection would not smile on such a village—the men would not cooperate in the hunt, women would not gather, food would not be distributed efficiently, and children would go hungry. Culture then functions in part to safeguard individuals' reproductive and economic interests in a society, in return for an individual's contribution to the overall societal well-being.

Other aspects of culture function to organize an individual's work and production in a society for the benefit of the whole group. They promote a strong bond and sense of identity with the group, melding an individual's drive toward personal preservation into a feeling of group responsibility. A society held together by strong cultural glue will be at a competitive advantage in confronting the elements as well as in any competition with other groups. Traits held in common in a group will enhance their common identity and thus underline their common cause, which at the base is what natural selection acts on—survival and reproduction. Such cultural traits as hair style, body ornamentation, jewelry, and clothing may seem far removed from, if not totally irrelevant to, natural selection. Indeed they may vary whimsically, but that very variation defines the group and pulls it together in an intimate way that is only matched by ties of kinship among members of the group. For a social species like humans, whose survival depends on the shared benefits of cooperation, such group solidarity is of tremendous adaptive value. When an environmental or external threat such as a famine, a flood, or an attack by a hostile neighboring group occurs, the culture can depend on its members to defend it.

But defend it with what? Clearly people can go out and physically fight with individuals of an invading group or other animals who want to displace them from their watering place, occupy their cave, or steal away their children. But how do you fight a famine or a flood? Members of the group will do their best to deal with the consequences of such disasters, but prevention would be preferable. Culture solves this problem with what we in English call "magic," but in many societies it is closer to what we call "religion."

As humans evolved greater intelligence and more associative properties of their brains, they began to understand that they had no real control over many of the causes and effects in the environment that could be dangerous or even fatal to them. But in the face of natural disaster it was essential to carry on as if they did. A group that gave up was sure to starve, but one that kept plugging might make it. A group that had exhausted all its obvious logical options may have hit upon magic as the only hope and it may have carried them through until things got better. Those groups whose magic and religion fostered the strongest "belief" were at an advantage, unless of course things never got any better and they died. But if this evolutionary model is correct, then in general a strong belief in group identity and the group's magic was fundamental to the cultural adaptation and evolutionary success of Pleistocene humans.

Magic evolved not only to "fix" things that went wrong but its power could also be harnessed to ensure beneficial outcomes in the future. Today some of us in the West retain a reverential awe for this magical power of "religion," but most of us put our firmest belief in our ability to control the environment by what we term "science." Of course there are many important and fundamental differences between religion and science, but in the sense of their functional importance in cultural adaptation to the environment, they are the same.

Language serves as the cultural basis for retaining and communicating accurate knowledge about the environment. Its structure permitted organization of knowledge into myths, epics, and eventually books to which people could refer and remember. Language, for example, allows specific labels to be placed on different species of animals and plants, naming of locations where they are found, description of their characteristics, and discussion of their current dispositions in relation to the season and weather. Transmitting knowledge about the environment was likely a centrally important purpose for complex language to evolve. Still today our most basic of conversational topics is, of course, the weather.

Even if storage and retrieval of information about the environment was the primary function of language, components of culture are all connected and they can affect each other. Advanced culture would be impossible without language, which is used by societies to express kinship relationships, to communicate what is correct and incorrect behavior, to pass on cultural information to the young, to entertain, and to mark major events in the life of the group. Language became a vehicle through which the many aspects of cultural adaptation were expressed. Origin myths, for example, summarize linguistically and symbolically humanity's relationship with the cosmos, imply an ideal way to interact with that environment, and explain apparent illogical or inconsistent phenomena.

Language is also a reflection of the action of natural selection on the human species in creating the human capacity for culture. The evolved human brain carries the developmental map for language, an attribute of language that linguists call "deep structure." Despite the fact that a specific culture has to supply the vocabulary, language ability appears spontaneously as a child grows up. There is thus a fundamental, if still unclear, relationship between biological natural selection and the ability to learn and use culture as exemplified by language.

Language plays a very important cultural role in defining the boundaries of a group and its identity—what is "we" and what is "they." "Accent," for example, is an important characteristic of a cultural grouping or subunit. It is determined by the age of puberty, after which learning a new language without a "foreign" intonation is virtually impossible. Throughout most of human evolution there was a clear-cut synonymy of language, the cultural unit, and the biological population unit. Thus, people who talked differently from you also dressed and built their houses differently, and even were usually somewhat physically different, with a different hue of skin, an odd shape of nose, or unusual form of hair. The proliferation of advanced culture toward the end of the Pleistocene witnessed an intensification in the intragroup identities and xenophobic interactions between groups while increasing the effectiveness of culture as an adaptive mechanism to the environment.

A species such as *Homo sapiens* in the end of the Pleistocene that became divided by culture into increasingly distinct subunits would be expected to have had responded to natural selection differently than their ancestors. Indeed it is likely that at this time such valued human traits as altruism and heroism evolved, as well as such despised but nevertheless pan-human traits as racism and genocide. These are all behaviors that fundamentally involve group identity and are strongly supported by culture. These are unusual behaviors. One is a selfless feeling that gives rise to a heroic act that benefits the group but can injure, maim, or even kill the individual who carries it out. The other feeling is one of extreme aggression based on nothing more than an individual's membership in a different group, be it a biological unit (a "race") or a cultural or linguistic unit.

When carried to extremes these feelings can lead to physical injury and killing of members of one group by members of another, an all-too-familiar and all-too-recent phenomenon termed *genocide.* How were these very human behaviors adaptive and how did natural selection produce them?

SOCIOBIOLOGY AND THE EVOLUTION OF ALTRUISM AND RACISM

Sociobiology, that integrative field of biology, anthropology, and psychology that attempts to explain behavior in social groups based on shared genes, has made major contributions to an understanding of human cultural evolution. Sociobiology started with the observation that in social insects whole segments of the population will give up their reproductive function and become drones, workers, and nursemaids to the queen. This unusual evolutionary situation was made understandable only by the realization that the unique *haplodiploid* genetic system of these insects meant that a female nursemaid insect would end up passing more of her genetic material to the next generation by helping her sister, the queen, rear her eggs rather than by having eggs herself. Despite the fact that our genetic system is vastly different in this regard from social insects, the analogy between bees committing genetic suicide and men marching off to war, leaving their mates and offspring to fend for themselves while they expose themselves to death, was a little too much for many sociobiologists to resist. Human sociobiology, sometimes also called evolutionary psychology, is now a burgeoning field.

One of the social phenomena that sociobiology attempts to explain is altruism—doing something potentially or actually detrimental to one's own good for the benefit of others. The more altruistic a culture can make its members, the more effective their all-out contribution will be to the well-being of the population, just so long as the society can ensure that nonaltruistic "cheaters" do not benefit at the expense of the altruists. Humans are indeed generally altruistic, but only when two social conditions are met: There must be a vital need or threat to the group and there must be a strong sense of solidarity within the group.

For much of normal functioning of a social group, altruism is not required. In fact, it would likely get in the way of many everyday activities and would unnecessarily expose individuals to danger. For example, there is little selective advantage in an individual forgoing eating to give food to other members of the group if there is no shortage of food, and throwing oneself in front of a passing cave bear intent on pillaging a nearby honey tree makes no sense. Thus, altruism is an adaptive mode appropriate for extreme situations.

We do not have any firm evidence for when altruism may have first evolved in human society or when it may have become most highly developed, but it is likely that the extreme environmental challenges of the latter Pleistocene would have put a premium on this behavior. It is certain, however, that altruism in humans is a much more conscious and volitional act than the altruism of social insects, depending on feelings of social solidarity and perceived threat. How then do we explain the genetics of its evolution?

Sociobiological theory holds that natural selection would favor the evolution of altruism only when an altruistic individual's genes are passed on to the next generation. Altruism only evolves when the benefit to an individual's genes, or copies of those same genes carried in other members of the group, outweighs the cost of the altruistic act. Stated less obtusely, individuals willing to go the limit for the good of the group will only evolve when they can help close relatives survive with future reproductive potential.

The statistics of genetic relatedness of sociobiological theory explains how altruism evolved in human groups. Early humans obviously did not have any way to calculate the genetic relatedness of their relatives in the group for which they were about to go into battle with a sabertooth. They were either altruistic or not, and natural selection acted accordingly. In mammals, an individual shares with its siblings and parents an average of one half of all the variable genetic information in the population (much of a species' genetic composition is already shared anyway, accounting for similarities in the species' anatomy and behavior). One quarter of the genes are shared with first cousins and grandparents. One eighth of the genes are shared with second cousins and great grandparents, and so forth. It is easy to see that in a group of twenty-five to fifty hominids, altruistic individuals, even if their mate was entirely unrelated to them, would need only one parent (sharing one half of their genes), one offspring (also sharing one half of their genes), and a smattering of cousins and grandparents (sharing one fourth and less of their genes) to easily outweigh their contributions of their own genes to the next generation, assuming the group survived. It made genetic sense to do things potentially injurious to oneself for the benefit of the group because groups throughout human evolution were composed mostly of related individuals. Altruism was bred into us by natural selection because we have always lived surrounded by extended families. Culture evolved in this context, building on the bonds of kinship and altruism that were already there.

The relatedness of most members of a culture throughout much of the course of human evolution helps to explain why all cultures are so self-centered—that is, why they always believe themselves to be the pinnacle of creation, that their way is the only right way, and that all others are less than human. There are lessons here for our present condition and future evolution (see Chapter 10). Cultures are basically ethnocentric and xenophobic. Although we may consider these traits less than admirable in today's ideal global community, the fact is that they were of extreme selective importance in human evolution. Absolute belief in the essential correctness of one's group and its cultural identity was critical for the survival and reproduction of the group. We have evolved to feel passionately about a few people and a few treasured ideals, and culture provides us the trappings to communicate and act out these feelings. The external manifestations of culture, such as ways of dress, hair styles, ceremonies, and ways of constructing houses, undergird a unity of purpose and a common response to challenges—an entity known as cultural solidarity. Cultural solidarity has been of extreme importance in meeting the environmental challenges posed to a tropical primate that found itself surrounded by the ice of the late Pleistocene.

The dark side of cultural solidarity appears when unified group behavior is turned toward other groups. This is a powerfully destructive aspect of culture and accounts for aggressive feelings that we can term *racism*. When human groups act out their racist attitudes they can become genocidal, attempting to destroy groups that they believe are different from themselves and that throughout most of human evolution would have competed for, rather than shared, environmental resources. Like altruism, racism is an attribute of cultural adaptation of extreme situations, coming to the fore when environmental conditions are the most extreme and life is the hardest. As such, it likely appeared in human evolution most full-blown at the end of the Pleistocene, during climatic episodes of extreme cold and dry conditions when food, water, and shelter were scarce.

THE EXPLOSION OF CULTURE
IN THE UPPER PALEOLITHIC

Archaeologists divide the European "Stone Age" or Paleolithic into *lower, middle,* and *upper* portions. The Lower Paleolithic (called the "Early Stone Age" by African and Asian archaeologists) extends back a couple of million years. It is typified by simple flake tools, large chopping tools, and relatively crude hand axes for the most part. The middle Paleolithic of Europe and contiguous areas of Asia and Africa is largely associated with Neandertals and is typified by long flakes and big blades. There are some broad regional differences in these earlier phases of archaeological assemblages, but nothing to compare with what happens when the Upper Paleolithic comes in. The Upper Paleolithic is generally associated with anatomically *Homo sapiens* and dates to after 32,000 years ago, the date of replacement of Neandertals by *Homo sapiens sapiens.*

It is not hyperbole to describe the rapid cultural evolution that occurred at the Upper Paleolithic as an "explosion." Nothing like it had happened before and nothing like it has happened since, unless we want to consider the continuing reinvention of modern culture as an ongoing explosive reaction that started at this time. Unfortunately we will never know all the many and varied responses to the environment that occurred in human groups throughout the world as the last glacial maximum advanced. But there is nothing speculative about the dozens, if not hundreds, of local stone-tool traditions that began to appear at this time. Archaeologists have been divided on whether these clearly different tool assemblages represent different cultures and different groups of people or merely different adaptations within the same culture employed by the same group of people under different environmental conditions and perhaps at different times. Although it is impossible to interpret the archaeological record uniformly in this regard, the length of time and regional variety over which the record is spread would tend to suggest that we are sampling in general different groups and different cultures throughout much of this record.

The archaeological record of stone tools is but the most durable of the aspects of the cultural burgeonings that we detect during the ice age. We surmise that because all aspects of culture are tied together today, they were similarly interwoven in the past. A change in a group's tool kit then would have been reflected in many aspects of the culture less prone to being preserved in the archaeological record, as the hypothetical example at the beginning of this chapter showed. There is more to this conclusion than analogy with modern culture, however. We have two aspects of the archaeological record, art and music, that support this interpretation.

No indication of representational art is known prior to the late Pleistocene, when climate change was putting more and more pressure on human groups to adapt to progressively harsher environmental conditions. The earliest example of art is a crudely carved figurine of a pregnant woman excavated in Israel by Naama Goren-Inbar and dated to 100,000 years ago. The first examples of cave and rock art paintings in Europe and Africa date to at least 75,000 years later and correspond closely in time to the onset of the glacial maximum, which culminated at about 17,000 years ago. A number of "Venus figurines," similar to the early Israeli discovery, also date to this same period.

Modern anthropologists cannot possibly divine in detail what the meaning of this earliest art was. But its significance in general terms cannot be in doubt. Unlike modern Western art, art in traditional societies exists in a much closer equilibrium with the rest of the culture that nurtures it. A Venus figurine was not a purely aesthetic piece of art; it was embued with magic and symbolic significance. Its exact meaning to the culture that produced it escapes us now but we can be certain from its subject that the Paleolithic artist was somehow placing a high value on female fecundity, reproduction, and a well-fed state. No ideal of the svelte here. A straight-line correlation of high cultural value and reproductive success is hard to avoid. Culture furthers the imperative of biological natural selection but at the same time enhances the differences and hence the differential reproduction of separate human groups.

Most of the cave and rock art paintings that have survived are of animals—not animals that were admired for their beauty or swiftness alone but animals that people could eat and that they had to catch and dispatch. We can surmise from the context of these paintings that they represent a close cause-and-effect relationship between powerful cultural magic and successful hunting in an extremely harsh environment. The paintings are located in the innermost recesses of caves, not next to the hearth where one might gaze fondly on a pleasant scene. Ethnologists have termed the type of magic probably practiced with the use of representational art as *sympathetic magic*—what you do to the picture will translate into what happens in real life. The many shafts piercing the animals of the cave walls support this interpretation.

One idea that ethnologists have entertained about the origins of magic and religion relates these cultural attributes to dreams. People enter into another world sometimes when they sleep and they see and experience things that are part real and part unreal. We in the modern Western world have cultures that largely emphasize the unreal part of dreams. Many traditional cultures have chosen to look at dreams as primarily real. One could logically conclude with this latter assumption that if one sees people and places in a dream that are real, then everything seen in a dream is or could be real too. If everything in a dream is real, then people and things have spirits that can leave their bodies and enter your head while you sleep—all perfectly logical. Art may have evolved as a way to capture the spirits of those animals that you needed to hunt or to control the mysterious forces of human reproduction. The intense darkness of an inner cave would have simulated the darkness of night and the sensation of sleep when you closed your eyes, and thus may have represented an entry into the spirit world for Paleolithic people.

The Upper Paleolithic was also the first time that evidence of music appears. Simple flutes carved from the long bones of animals have been found among the bone refuse left over from innumerable meals. Again, how these instruments were used and what ancient melodies they may have played will probably never be known. But an interesting bit of detective work by a group of European archaeologists in several cave sites provides some indication of their possible context. These researchers were puzzled by simple blobs of paint smeared in isolation at certain points of the inner recesses of caves where art was found. They assumed that they were some sort of trail-marking system or a method for locating ceremonial sites until they discovered that every point so marked was a resonance point for sound. With sophisticated equipment they reproduced the frequencies characteristic of the surviving bone flutes and found that at each point a harmonic resonance was produced. Music, which can certainly be magical in its transformation of the human spirit and is also very closely tied to cultural tradition, may have been born as an early way to contact the spirit world, which answered back at those Paleolithic listening posts in the depths of dark caves. Music then would have evolved as part of the overall cultural adaptation to ensuring reproduction and survival in an increasingly hostile environment.

GEOGRAPHICAL EXPANSION
OF THE ADVANCED CULTURE BEARERS

Paleolithic artists never sculpted human beings, except for the mother goddess figurines, and they virtually never depicted any people in their cave art. This is a most curious fact and it is undoubtedly the main difference between Paleolithic art and our historic art traditions. Art in the Pleistocene existed as part of a utilitarian adaptation to magically control the environment, and if people were not subjects of Paleolithic art, then the clear implication is that people were not a major unpredictable and uncontrollable concern. Perhaps Paleolithic people knew how to deal effectively with other groups of people around them, because judging from the cave art, they were not particularly worried about them. Magic and religion are generally not invoked when people are secure in the outcome of a future event or if they are not threatened by it. How then do we reconcile this apparent lack of the use of magical power against other humans with the conception of culture as intrinsically ethnocentric, xenophobic, and potentially genocidal?

In an ecological sense, cultural differentiation may have acted as an effective spacing mechanism for human populations during the latter part of the Pleistocene. Cultures did not come into conflict with each other so much as they mutually avoided each other.

Each culture repelled another, much the same as positively charged ions in a solution repel each other. Each culture was (and still is) content to look upon itself as the only and best way to live, and to regard the rest of humanity as largely irrelevant. Using modern human cultures as a guide, only when environmental resources such as food, water, and living space became limiting did actual conflict take place.

Thomas Malthus was correct in assessing the intrinsic rate of population growth to be much faster than the rate of increase of productivity of food. We can assume that the expansion of population size regularly reached the carrying capacity of a region when other checks on population growth such as death from injuries, disease, or old age were lessened. This situation undoubtedly led to intergroup conflict in some cases. In other cases the natural avoidance of one cultural group for another would have tended to spread groups out into all habitable areas, including some areas that had previously not been inhabited by humans.

The spacing mechanism of culture, coupled with the intrinsic tendency to increase population size, would have tended to push human population expansion at the edges of its distribution. Two main types of geographic barriers would have stood in the way of late Pleistocene geographical expansion of human groups—ice and water, both different states of the same H_2O molecule.

Glacial ice was an environment inhospitable to most plant life as well as to most animal life. Large mammals were able to live at the edge of the ice, however, making a living from grazing and browsing on small hardy plants, such as lichens, and preying on other species that lived on land and in the sea next to the ice. Humans were able to live in these very extreme environments by preying on the animals that had adapted to these glacial conditions. Culture, with its flexibility of adaptation and its ability to effectively coordinate the activities of a large body of people, was the key element in making this adaptation possible.

Humans were and are a terrestrial species unable to cross large bodies of water by swimming. Swimming is a metabolically expensive activity for people, there is nothing to eat or drink in the sea, and humans are subject to rapid loss of body heat since they have no insulating fat or hairy coat. During the Pleistocene both environmental changes in the distribution of water and cultural means of traversing bodies of water allowed humans to colonize the rest of the habitable world. As glaciers advanced, more water was locked up in the relatively immobile solid form of water: ice. Sea level consequently went down around the world. The estimate of the amount of sea-level drop during the height of the last glaciation is 120 meters, or about 350 feet. This amount was easily sufficient to connect the British Isles with the European mainland, northeastern Africa with Arabia, Java with the Malay Peninsula, Japan with mainland Asia, and—most importantly from a standpoint of the expansion of human territory—northeastern Siberia with Alaska. People walked across these "land bridges," where previously there had been deep water. Most of these places had seen human beings before, but the late Pleistocene was the first time that humans made it across to the New World, to North and South America.

Other land areas of the world remained unconnected to the Old World landmass by the late Pleistocene lowering of sea level. The island continent of Australia, separated by deep water from Southeast Asia, remained isolated even during the maximum lowering of sea level in the terminal Pleistocene. Australia's fauna and flora, delimited from the mainland Asian biota by "Wallace's Line" running along the Straits of Molucca, kept to itself. Only the incursion of one large placental mammal, *Homo sapiens sapiens,* disturbed a domain otherwise ruled over by the primitive pouched mammals, the marsupials. Evidence indicates that people first appeared in Australia at about 40,000 years ago. Since there was no land connection, these people must have gotten to Australia by boat. This cultural innovation made the entire world potentially habitable by humans. Boats had to have been utilized also in the peopling of the Pacific Islands, but available evidence indicates that these population dispersals happened much later, probably at about the same time as the peopling of the Americas.

Paleoecology and the First Americans

When Columbus first encountered human beings in the "New World," he and other Europeans made both a geographical error and an anthropological mistake in naming them "Indians." Columbus thought that when he arrived at Hispaniola he had reached some islands off the mainland of the Indian subcontinent. An oft-neglected support for his conclusion was the clearly Asian features and physiognomy of the inhabitants. Columbus took several Arawak "Indians" back with him to the Spanish court to bolster his claims. Trading had been underway with India and the Far East since the times of Marco Polo and the appearance of Asians was well appreciated by many Europeans. Most had to admit that Columbus's claim that they were Asian looked reasonable.

After further exploration during the early sixteenth century exposed the geographical error of Columbus's first pronouncements, the anthropological component of his original conclusions remained a mystery. The inhabitants of the New World continued to be called "Indians" for want of any better term and because no one had any idea where they had really come from. Despite some speculations by the church that the American Indians might represent some of the lost tribes of Israel, it was more than 250 years later that Thomas Jefferson reopened the question of the origins of the original inhabitants of North America. He suggested in *Notes on Virginia* that American Indian languages bore resemblances to the languages of Asia, and thus they had probably originated from this source. Despite some other ideas that people may have island-hopped from Australia around Antarctica to reach South America, had rafted across the Pacific from Japan to the west coast of South America, or had boated from West Africa to Mexico, the Asian origin of Americans has been the working hypothesis ever since.

Archaeologists in the New World have debated for nigh on a century about the earliest evidence for humans in the Americas. Fossil human bones have remained elusive, but the available archaeological evidence has been cited in support of either one of two positions—the early colonization or the late colonization of the Americas. The early colonization appears on general principles to be quite reasonable. There was clearly a land bridge across what is now the Bering Strait at several times in the Pleistocene and we know that as early as *Homo erectus* there were early human populations in the northern part of Asia. There is no a priori paleogeographical reason that populations could not have migrated across the Bering land bridge at low stands of world sea level, at 70,000, 90,000, or 110,000 years ago, or even during a long period of time extending from 130,000 to 175,000 years ago. But generations of archaeologists have failed to provide incontrovertible evidence of any human presence in the New World prior to the maximum of the last galciation, at circa 17,000 years ago. The late-colonizing school therefore still holds sway.

The relatively late peopling of the Americas makes some evolutionary sense of the fact that all populations in North and South America look so similar to each other. If they all came from a relatively recent migratory event between some 12,000 and 17,000 years ago, there has not been sufficient time to cause significant differentiation of human populations in various areas. For example, the New World is the one major region of the world where human beings near the equator do not have darkly pigmented skin. The ancestors of these populations would have come from the far north and thus would have had more lightly pigmented skin. Natural selection would not have had time to change skin color in response to the deleterious effects of incresed sunlight in such a recently arrived population.

Another observation that tends to support the position of the late-colonization school is archaeological. Almost all the earliest sites date to around 15,000 years or later, and interestingly they are found all over North and South America at virtually the same time. This near-synchrony in dates of the earliest archaeological sites suggests that colonization was rapid. One would expect the earliest evidences of migration to have been found in the northwestern part of North America, but the earliest dates so far reported are from several cave sites in South America. One of the most convincing recently discovered sites for the earliest human presence in North America is Saltville, located in the middle Appalachian Mountains of southwestern Virginia, on the *eastern* side of the continent. The most reasonable way to interpret these data is to suggest that the migration of latest Pleistocene humans into North America was rapid and unimpeded by prior human presence.

Genetic studies have supported a late Pleistocene migration of a "founder" population with "paleo-Asian" affinities, represented today in similarities with populations living on the Indian subcontinent. A later wave of migration into the Americas accounts for the genetic similarities of "Eskimos" with extant native populations in Siberia.

If this scenario of the first arrival of Asian immigrants in the Americas is correct, then the movement of these populations represents a "rebound effect" from the ameliorating conditions following the glacial maximum at about 17,000 years ago. Cultural adaptations that had pulled people through the worst glacial conditions of the Pleistocene allowed them to expand their range into two continents that had a glacial gateway that had never been breached by hominids before. Recent pollen evidence shows that the Bering land bridge would have been a stark and barren tundra 15,000 years ago. A human group making this trek would have had to have been well adapted to surviving in these conditions. Not surprisingly, the Saltville site, dated at 15,000 years ago, has the remains of a mastodon kill arranged ritualistically. Culture had been the most important adaptive difference that had brought late Pleistocene humans from Asia, where their predecessors had been turned back. Other cultural advances such as boats and ways of extracting food resources from the sea played an important role in allowing this dispersal across the land bridge.

OUR ANCESTORS, OURSELVES:
CULTURAL AND ECOLOGICAL ADAPTATIONS
IN THE MODERN WORLD

Since our individual lives are so infinitesimally short, it is easy to forget that we are still essentially in the Pleistocene. The glaciers will come again in a few short millennia. And rewinding just a few frames of our recent evolutionary history puts us squarely back around a blazing fire in an ice-age cave or in a camp beside a dying stream at the edge of an encroaching tropical desert. Culture allowed us to survive those challenges. Will it be up to the challenges of the future? Will humans even be around long enough to find out?

The ecological interplay of culture, with all its technological sophistication—or what passes for that in our minds—and the environment has been a topic of central social concern in America for over three decades now. The general opinion is that we are ruining our environment, and that we should stop, for the sake of the environment. As usual, however, when humans have a concept, they are always at the center of it. In fact, the "environment" as a whole does not revolve around the human species, and it will survive just fine, minus us and a few other species that we may take with us to extinction. The pollution that we have dumped into the waters of the earth, buried in the ground, and emitted into the air will one day constitute only very thin layers in the geological strata of the earth. It is we, and it is *our* environment, that we are ruining. The fouling of our nest and the destruction of our ecological niche are the subjects of the next chapter. How can a species that has existed for some 500,000 years find itself now so close to its ecological tolerance limits? What has gone awry in our ecological adaptation and what can we do about it?

10

Future Human Evolution, Overpopulation, Global Warming, Pollution, and Our Ecological Survival

The preceding chapters have presented a series of hypothetical formulations that attempt to account for the major steps in hominid evolution. This is a long and twisting narrative, but the consistent theme has been that ecological and environmental changes have had significant and indeed formative effects on human existence and evolution. Ecological change has been the "prime mover" of human evolution. The major stages of this evolutionary history have been known for some time and they have been confirmed and refined by many intersecting lines of evidence—anatomical, paleontological, archaeological, and geological. With the burgeoning and increasingly accurate data about the earth's past environments, it has been possible to tie in ecology to the major steps in human evolution. We have seen that the parameters of human existence have been molded over time by environmental change and natural selection. The intensely interdisciplinary approach of paleoanthropology has provided cross-checks on these hypotheses from different disciplines. It is reasonable then to draw some conclusions from this history and to apply them to current concerns.

FUTUROLOGY AND THE TIME SCALE OF HUMAN EVOLUTION

There are scholars who make a study of the future. Many of them, called "futurologists," are political scientists and economists, whose concerns are "long-term" financial and global stability, resources, and growth. In the several years that I worked in Washington, D.C., I came across several of them, but I was amazed at the time scales with which they dealt. "Long-term" for them tends to mean simply "as far as we can extrapolate linear trends." The lines on their graph paper run out only about fifty years into the twenty-first century. One suspects that "long term" translates as an optimistic assessment of a political term in office or a career as a loan officer in the World Bank. Thus, at first glance there may appear little that a perspective as broad and as long as paleoanthropology (or perhaps as coarse-grained and irrelevantly ancient, if one talks to a futurologist) can contribute.

Anthropology's perspective on human evolution, coupled with an ecological perspective on human adaptation to changing conditions, has a tremendous potential to inform and guide public policy and planning for society's future. It will clearly not yield times and dates for future evolutionary events to transpire but it can most assuredly point to disequilibrium conditions that will, certainly, change. It will also tell us in which directions human behavior is likely to be directed, given a certain set of environmental conditions. In some cases the knowledge of human evolution will guide us toward a potential solution, and in other cases it can inform us that past attempted solutions have not worked. This chapter will examine some of the major problem areas of human–environmental interaction.

Present and Future Human Evolution

There is a widespread belief that in the modern world human evolution has ended, that it is only a thing of the past. The origin of this idea is probably the belief that modern society is so powerfully controlling of human destiny that it has overcome all constraints of nature. This is incorrect. So long as there are birth and death, there will be varying degrees of reproductive success among people, differential survival of offspring, different degrees of access to food and other resources for life, and a range of variation in the human population. Natural selection will continue to work. As we have seen, human culture, the basis for the multifarious societies in which human beings live and reproduce, evolved out of the natural realm and is a product of natural selection. Human culture is a powerful adaptive force in its own right, but only if it ultimately serves natural selection. What nature gives it can take away.

Our fragile societies continue to exist at the whim of nature, and should culture fail to be up to the challenge of future environmental changes, then natural selection would simply come up with a different solution. As Michael Crichton in the fictional *Jurassic Park* has one of his characters intone, "nature will find a way." Nature's solution may or may not be a good deal for the human species. We may become extinct, a "solution" that from our perspective is hardly optimal. We thus may benefit from using our brains, our most powerful adaptation to the environment, to figure out the rules for the continued coexistence of ourselves and the world around us.

It is important to frame the problem of human–environmental

interaction in appropriately humble terms. We may destroy the "world around us" for ourselves and we may even take down several species with us to extinction, but it is not within human power to literally destroy the earth. We are far too puny to do that. Our most powerful nuclear weapons would only leave a few minuscule pock mocks in the earth's crust where our major cities had been, and within a few centuries even these would be erased by erosion. Humans and their highly touted culture would be gone, but other species—bacteria, plants, insects, and even probably most vertebrates—would remain and continue to evolve. The earth would survive just fine. So we are not discussing altruistic "save the earth" campaigns. We are discussing "save the earth for us."

THE GLOBAL PROBLEM OF OVERPOPULATION

In one of the most important cross-disciplinary fertilizations in science, the ideas of economist Thomas Malthus gave naturalist Charles Darwin some of the working concepts for the theory of natural selection, published in Darwin's *Origin of Species* in 1859. Malthus's formulation was simple. Populations increase geometrically, graphing to an ever-increasing upward curve, while the food resources on which they depend increase only arithmetically, graphing out to an upward-sloping straight line. As population size increases and the lines on the graph diverge, the greater the disparity between the haves and the have-nots. Malthus could only see suffering, famine, disease, and war in the human future unless sexual continence could curb human population growth. Eventually there is an equilibrium point at which the rate of production of offspring equals the death rate, and it is at this point that Malthus's upward curve of population growth flattens out to the top part of an *S*. Malthus's formulation did not posit any absolute maximum size for population. Rather, he pointed out the critial relationship between population size and resources. Malthusian disaster can therefore occur not only when the population increases past a point at which resources can support it but can also strike when resources shrink relative to the number of people in a static population.

If world history is any guide, Malthus was essentially correct. Large-scale depopulation has occurred in the face of disease, such as the European "Black Death" of the fourteenth century when an estimated 75 million people, between one quarter and one half of the population, died. The Black Death (bubonic plague) was caused by a bacterium, *Yersinia pestis,* that was spread to humans by fleas that had bitten infected rats, and also by direct transmission from individuals in whom the infection had spread to the lungs. High population density in cities greatly facilitated the spread of the disease, although an environmentally induced increase in the rodent populations in which the disease was endemic has been suggested as the proximate cause of the epidemic.

An earlier epidemic of plague in the Middle East, Europe, and Asia in the sixth century is estimated to have wiped out 100 million people, perhaps one fourth of the world's population at that time. Smallpox, an introduced European disease, decimated the native American Indian population in eastern North America during the eighteenth century. Beyond the reach of history but still recorded in our genes is the terrible toll exacted when malaria became endemic to Africa, leading to protection by the sickle cell trait. A similar defense evolved against cholera in European populations, leading to a genetically mediated immunity to cholera toxin by the cystic fibrosis gene.*

*Individuals infected with the malarial parasite are protected from the disease by sickling of the red blood cells, which are removed from the blood by the immune system along with the parasites. But if all the individual's cells are sickled, then sickle cell anemia results, a fatal genetic disease if left untreated. There is thus a "balanced polymorphism" in African populations between death from malaria and death from sickle cell anemia, a lasting legacy of the natural selection resulting from this disease. A similar relationship exists in European populations between cholera and cystic fibrosis—individuals who have one gene for cystic fibrosis do not have this genetic disease and are protected from the infectious disease, cholera, whereas individuals with two genes are afflicted with cystic fibrosis.

Increase in population numbers and greater population density are both implicated in the spread of disease. Malaria, for example, became widepread in Africa only after agriculture cleared the land and villages increased in size, both of which established standing pools of water, which are breeding sites for mosquitoes, the vectors of the disease. Cholera is caused by an intestinal bacterium *(Vibrio cholerae)* that is spread through contamination of a common water source by feces. It is directly related to population density.

War, another Malthusian population check, has been an important force of depopulation. World War II led to the deaths of 15 to 20 million combatants and over 25 million noncombatants, not including the 6 million Jews who died in the Holocaust. If we think that the Nazi Holocaust was a one-time event, we need only look to the decades-long Hutu-Tutsi conflict in Rwanda or the "ethnic cleansing" campaigns of the Balkan civil wars, which have killed hundreds of thousands and displaced millions.

Overpopulation has a direct connection to war, usually expressed in terms of economic distress. Hitler's rise to power after the Depression on a platform of xenophobia and economic stability would have been impossible without the widespread feeling in Germany that there were too many people and too few resources. Rwanda's population density is the highest in Africa and competition for arable land is intense. The economic distress of Bosnia, Croatia, and Serbia after the collapse of the Soviet and Eastern European socialist economies is the proximate cause of the civil wars that have riddled the region.

Famine also occurs throughout the world with regularity. The failure of the potato crop in Ireland in 1841 led to widespread starvation and the death of some 1.5 million people. Drought in Ethiopia in 1990–91 affected some 7 million people and resulted in the deaths of uncounted hundreds of thousands. Famine occurs when the population is already near the ecological carrying capacity of the land. With one environmental perturbation such as a fungus *(Phytophthora infectans)* affecting one plant (the potato) or a relatively low rainfall over a two-year period that does not allow herds on already overgrazed lands to survive, whole-scale starvation can result.

A curious fact connects all these sources of ecologically driven depopulation. "Group" membership defines their action and effects. Disease is a population phenomenon, affecting some groups and not others. War is always between "groups," and the side that wins gives up the lesser resources and the lesser number of people to the opposing side. And even such an apparently nonpartisan grim reaper as famine somehow in humans shows favoritism, because cultural factors still determine how the few remaining resources are distributed. This way of looking at depopulation as a group-related phenomenon is key to understanding the human evolutionary response to the problems associated with overpopulation.

In times of crisis our evolutionary tendency is to pull back to a defense of our most inclusive and smallest group—the ancient kin-related extended family/tribal nucleus. Natural selection has acted on this basic group. It is thus entirely "natural" from an evolutionary and ecological standpoint for human beings to cohere to "their" nuclear group. Natural selection and cultural evolution have both acted over eons of time to achieve this response, which up to now has been effective.

A nuclear group response to ecological threat is no longer adaptive in the interglobal world of today where challenges of global scale confront us. But in the past this was exactly the level of response required. When environmental resources were scarce and survival seemed tentative, the imperative from natural selection was to have as many offspring as possible, in the hope that some of them might survive. Evolutionary ecologists called this *r selection* (see Chapter 1). Today people in poverty in the United States and throughout the world are playing out this evolutionary strategy. The highest birth rates are consistently found in those segments of society and in those countries that are the least affluent, that have the poorest housing, that lack access to adequate food, and that are the least educated. Populations that badly need to limit their growth so that limited resources can be shared more effectively are thus caught in a bind with an ancient human evolutionary ecological adaptive mechanism.

Legislating population growth by governments has proved historically to have been of limited success. When France wanted to increase its population in the nineteenth century for the good of the state, it banned both abortion and contraception. But birth rates did not go up as expected because couples reportedly practiced *coitus interruptus* and illegal abortions became commonplace. Modern China, on the other hand, acted to curb population growth by legislating that couples may have only one child or face stiff fines. There is evidence that this policy has been effective in reducing the rate of population growth in China, but at a cost to personal freedom of choice that many in the West would abhor.

One factor, however, does consistently lower the birth rate in populations. Paradoxically, this factor is increasing family income, which correlates with equalizing of male and female incomes, better housing, and a higher level of education. One economist has noted, "[i]t is as though people will only limit the number of their children when they see some chance of rising out of poverty." In ecological and evolutionary terms, humans will shift over to a K-selection strategy when they are secure in their predictions of adequate food and resources for their family group. Without that security they will tend to remain in an r-selection survival mode of reproduction.

The threat posed by the ever-increasing world population means in Malthusian terms increased suffering, more widespread disease, war, and famine, not only for the affected population directly but for all of us indirectly as environmental degradation, pollution, disease, lowered economic productivity, and conflict can now easily spread globally. World population now stands at 5.3 billion, and one estimate suggests that if birth rates are brought into line with death rates, world population will level off at between 8 and 9 billion by the year 2075. But another futurological study by the United Nations suggests much higher numbers—between 11.3 and 14 billion people—by the end of the twenty-first century.

Decreasing birth rates worldwide, rather than waiting for Malthusian population checks, is obviously the best solution, and we have seen that governmental actions have been only marginally successful in legislating compliance. Any solution that is going to be effective in the long term has to be in accordance with our basic adaptive mechanisms evolved over millions of years. We should seek therefore to shift populations under duress from an r-selective mode of reproduction to a K-selected mode, and thereby bring down the birth rate in countries where it far outbalances the death rate. This solution requires that "rich" countries reorient their international aid efforts toward preventive aid, rather than after-the-fact disaster aid. The effective distribution of technology, education, food, and medicine from the developed world to the developing world, as well as including these countries in the economic loop of the developed countries, will contribute to this goal.

Even if this preventive approach to solving the world population problem will not cost any more money than the existing programs of after-the-fact disaster aid, it will still be a difficult program to get enacted into legislation. Legislatures answer to the "groups" that elect them, and these groups are primarily concerned with their "group" and no one else. This response is in a sense an evolutionary legacy, born of millennia of adaptation to small, closely knit groups. The human ecosystem of today and the future, however, are vastly different.

We now live and interact in a world with thousands, millions, and even billions of other people—numbers that the human mind cannot intuitively comprehend. We do not and cannot know even a small proportion of the people sharing the planet with us, but their actions will as a group affect our lives as much as those of our closely related kin and everyday "group" associates. Intellectually, it is clear that a pluralistic global society will have to evolve, and that in some manner "they" are going to have to be defined as "we" before any real progress toward solving the imbalance of world population and available resources can be made. In this endeavor both our variegated cultures and our biology are excess baggage. For better or for worse, we cannot leave this baggage behind; it must accompany us on our forward journey. But at least we can learn to deal with and work around the weak points of our adaptation. As world population has grown, some of these weak points have become obvious and of critical concern to our continued survival. The rest of this chapter will examine the most important of these malfunctioning aspects of our adaptation to the environment.

The Urban Human Zoo

Darwin pointed out over a century ago that in many respects, such as in the loss of our extrinsic ear muscles, human beings resemble domesticated animals. Except for a few adolescent party tricks, our extrinsic ear muscles are totally useless. They are like the floppy ears of sheep, goats, cows, many domestic dogs, and even some breeds of rabbits that lost the ability to respond to the slightest crackling of a twig at the approach of danger. Instead, humans protect domesticated animals from predators and at the same time become their primary selective agents. Darwin studied this and many other examples of artificial selection in order to gain insight into how natural selection worked. He thought that humans resembled domesticated animals because people had essentially domesticated themselves—by removing the usual forces of natural selection and replacing them with the forces of artificial selection. Today we would explain this self-domestication of humans as adaptation to a cultural ecological niche—humans having adapted to culture, which in turn adapts to the environment.

Zoologist Desmond Morris took Darwin's idea further. Morris observed that human beings resemble not so much domesticated animals as caged animals. In *The Human Zoo* he argued that humans murder members of their own species, have a heightened sexuality, are overly aggressive, become obese, and suffer from neuroses, much the same as wild animals confined in zoos. Implicit in Morris's message was that the urban habitats in which modern humankind finds itself today are not its aboriginal or optimum environments. If this is true, how did such a discordance ever come about?

Ever since the Neolithic Revolution some 10,000 years ago in Mesopotamia and the Nile Valley, when people started practicing agriculture, animal husbandry, and the control of permanent water sources, they began to congregate together in knots of population and closely spaced structures known as villages, towns, and cities. At first this adaptation allowed the coexistence of traditional hunter–gatherer life ways beyond the confines of the early cities and their fields. But as the Middle East was, and still is, a region of marginally adequate precipitation, progressive growth of population in the cities contributed to growing herds, overgrazing, loss of topsoil, deforestation, and killing off of the local wild animals. Soon there was no more place for hunters and gatherers.

Cities found permanent sources of water and sent out tentacles of trade that allowed them to survive and continue to grow even though they were surrounded by wasteland. Essential to their success was the growth of regional economies based on trade. Cities were able to attract the "peasants" from the hinterlands with the promise of a part of this wealth, a pattern of population movement that has continued to the present day in most countries of the world.

Accompanying people in this move into the city were generally lowered standards of living, a smaller living space, less varied diet, a greater chance of infectious disease, widespread prevalence of nutritional deficiencies, a slower growth rate for children, and a much higher chance of crime. We can accurately date the onset of urban living in the archaeological record by the decrease in people's body sizes and the increased incidence of disease, as indicated by their bones. With the many drawbacks to life in the city, why reasonable people would choose to move from the open countryside to such a place is mystifying.

The phenomenal growth of cities over the last 10,000 years attests to the extraordinary power of culture in directing and affecting human behavior. Cities with their concentrated populations become cauldrons of cultural activity, in turn spinning off economic productivity, political activity, and excitement. For a primate whose brain is prewired for cultural stimulation, a city becomes a mesmerizing population magnet. But, ecologically, cities are a bad idea.

The first urban environments with their increased population densities went along with increased concentrations of human waste and garbage. As cities have grown they have had to deal with these basic public sanitation problems, or else face extinction. But as anyone who has seen the havoc resulting from a weeklong garbage strike in New York can attest, it is a fragile "solution," and one without a long-term future. Modern cities deal with their waste with their water treatment plants, sanitation systems, garbage collection systems, and landfills—all solutions in the short term. But urban environments have been around long enough now to have created some long-term ecological problems from the wastes that they have generated.

Modern cities offer much poorer living conditions than non-urban environments. Because of the density of buildings there is a decrease in incident solar radiation and few open spaces with vegetation. Visibility is decreased and noise pollution is at its worst in cities. Horizontal wind speed is decreased in cities, tending to trap the noxious emissions of motor vehicles. Because cities also have numerous "heat islands," there is increased convection at the city center, which brings in even more pollution from the outskirts. Cities are hotter and more humid, with a tendency for more fog, cloudiness, and precipitation than the surrounding countryside. When particulate air pollution combines with fog to produce smog, a health hazard exists. In December 1952, for example,

London experienced a particularly lethal smog that claimed the lives of 3,500 to 4,000 inhabitants.

People, however, do not need to die or even to get sick to become convinced that many cities are not fit for human life. In many large cities around the world, those who can, leave, and move to the suburbs. They may still commute into the city for work, but their home environment outside the city has cleaner air, drinkable water, some open space with trees and grass, reliable public utilities, good schools, and a crime situation that is under control. Lionel Tiger in *The Pursuit of Pleasure* calls all these "evolutionary entitlements" of every human being, an innate part of our ancestral adaptation to which all of us should have free access. But what if you *can't* leave the city?

Urban riots. They happen when you're angry and frustrated at not getting what any normal human being should have to live. You're hot. You can't breathe. It would be great to have a place to go swimming. Somebody manages to turn on a fire hydrant in the street that cools everybody down. But then the cops come and turn it off. Suddenly it's too much, and the situation explodes. It doesn't seem to make any sense, but it happens, with predictable and terrible regularity.

Economists and news commentators shake their heads in bewilderment as the evening news shows the violence, looting, and burning in yet another city center. "Why," they ask, "do people burn their own neighborhoods, the businesses in which they work, and their schools?" The economist being interviewed by a newscaster cannot make any sense of it. Once he has mouthed the required platitudes about depressed economic outlooks for our cities, the need for greater education and self-help programs in the ghetto, and how the government needs to do something, he concludes that City X has just dug itself into a deeper hole. Then he is off-camera. As he leaves the studio, the economist think to himself, "I really don't understand it. These people act as if they are *owed* something, as if they *deserve* something. You only get things in our economic system if you *work* for them, and these people have obviously not worked hard enough. Instead of working within the system, they are actually trying to tear it down. Crazy."

But it's not. Illogical, yes. Economically unproductive, yes. But whole segments of a population do not go "crazy." Rioting is a primitive lashing out against the forced confinement in an urban habitat that does not meet even basic human needs. Perhaps Tiger is right in paraphrasing Thomas Jefferson. Perhaps there are "inalienable rights" to a decent human environment born of our evolutionary history. Perhaps that is what the rioters feel that they are owed but cannot put a name to. Try to explain that one to Jesse Helms.

FOULING THE NEST: POLLUTION AND DISEASE

Chimpanzees and gorillas build "nests" of vegetation every night in which to sleep. It initially surprised primatologists to observe that these close relatives of humans frequently defecate in their nests rather than leave them at night. Assuming that this was the primitive behavior, anthropologists have theorized about when and how humans became fastidious. An increased level of personal hygeine may have come about as soon as humans became more sedentary. Apes move their nests every night, so it hardly makes a lot of sense for them to put effort into keeping them clean. Humans, on the other hand, at some point developed a "home base," to use archaeologist Glynn Isaac's term. They would forage and hunt from this base camp and return there every day. It would make evolutionary sense to keep the home base, the immediate surroundings in which one was living, clean and relatively free of offal, which would attract scavengers, insects, and disease.

Most likely this behavior was not a conscious decision; things just "smelled bad" and were removed. Human hairlessness may have evolved at this same time as an evolutionary response to warding off skin parasites such as fleas and lice. These animals have a reproductive cycle of about two weeks and would be left far behind in abandoned ape nests, but they would have time to lay eggs and reproduce if humans were staying around in one place for two weeks or more. However it evolved, humans have a strong concept of their own personal and group space—their home base—and they keep it tidy.

The archaeological evidence for home bases is compelling and quite ancient. Tools and bone refuse at Koobi Fora, Kenya, and Olduvai, Tanzania, attest to repeated occupation of one site for relatively long periods of time. Humans thus added on a dimension of "personal space" that apes did not have. In addition to their "nest," which we in English call a "bed," there was a larger area that was kept clean and free of refuse. We might refer to this space by the familiar term "house" but it could also be an apartment, a cave, a hut, or a tent. At Olduvai site DK, some 1.5 million years old, there is the remnant of a stone circle that is widely believed to be the archaeological remains of the first structure (maybe a hut but also possibly a widebreak, a game hide, or something else). Outside this space there is a peripheral space that might or might not be considered part of the "house"—a "yard," perhaps. Most cultures also keep this space clean even though it is space that may be shared with other members of the group. Then there is the space "outside." Here garbage is thrown. I have a vivid image of a woman in a rural village in Zaire one morning sweeping the red dirt space outside her and her neighbors' huts clean of the slightest bit of litter, but she stopped sweeping at a razor-sharp line of demarcation with the surrounding luxuriant green foliage of the forest. This was her dividing line between "inside" and "outside" spaces.

I have been struck in cultures around the world at the varying degrees to which "insideness" and "outsideness" are applied. In Appalachia one's yard is definitely "outside" and becomes a dumping ground for old cars, washing machines, and assorted junk. In the Middle East, however, one's walled garden or courtyard is a sacrosanct space that is kept spotless. Step into the street, though, and one is apt to trip over an assortment of refuse, dead animals, and old tires. The public thoroughfares are considered "outside" and thus their cleanliness is paid no heed. Western Europeans, on the other hand, consider that urban areas are also to be kept clean and free from litter. Their definition of "outside" tends to mean "away from people." I have been amazed at my continental European colleagues in the field who nonchalantly throw an empty soft drink can onto the ground in the midst of the unspoiled splendor of the African savanna. They don't do it again if they see me picking up after them, because they had thought it didn't matter. Nobody lives there, nobody would ever see the can, and I don't live here, so what difference does it make? For 5.5 billion humans descended from nest-fouling apes it makes a tremendous difference. But we can learn to do better.

Pollution is basically an "inside–outside" problem. People define some place as culturally "outside" and consider it as appropriate for dumping waste. In the evolutionary past, when hominid population densities were low, as we saw in Chapter 6, and the global population only numbered a few hundred thousand to a few million, this time-honored approach worked fine. A relatively sedentary hominid group avoided close proximity to waste and thus kept the scavengers, insects, and disease at bay. Biodegradation took care of what was thrown "outside." But as humans have spread over the earth's surface and population has climbed up in a series of explosive bursts after the end of the Pleistocene, it has become more and more difficult to find a place that is "outside."

Love Canal is a place in upstate New York that a local government defined as "outside." Tons of toxic chemical waste were dumped in a landfill there. Years later the site was bulldozed, relandscaped, and turned into suburban homes by real estate developers. Love Canel was redefined culturally as "inside" but unfortunately the chemicals did not know that. They infiltrated the drinking water and food of the residents and made them sick, deformed their babies, and caused some of them to die of cancers.

Minamata Bay is in Japan, and because it was in the sea, a chemical company decided that it was "outside." The company disposed of large amounts of waste containing mercury in the bay. Fish living in the bay ingested the mercury. People eating the fish became poisoned with the excess mercury and, worst of all, pregnant women gave birth to terribly deformed and severely mentally impaired children. Doctors termed this new malformation "Minamata disease," but it was not really a disease; it was poisoning.

The attitude that "outside" could mean far away from people was tried out by the French government. It conducted atmospheric nuclear tests in the South Pacific, about as far from France as one can get. Unfortunately, other people do live there. The nuclear contamination in the form of an isotope of strontium, strontium 90, drifted over the Marshall Islands. Strontium 90 is absorbed into bone and stored there much the same as is calcium. So what appeared to be very small amounts of the contaminant in the environment and of no real concern became concentrated in people's bodies. As the radioactive strontium decayed and emitted DNA-damaging radiation, it caused the developing white blood cells in the bone marrow to begin dividing uncontrollably. This is known as leukemia. Marshall Islanders began dying of leukemia and suffering from skin cancer, where there had been a low incidence before.

The solution to pollution is to culturally define the entire earth as "inside." Then humans will have their inherited and culturally passed-down approach to environmental tidiness on the side of conservation. Our ancient hominid behavior of simply throwing things away so we do not see them anymore will no longer work. Unfortunately, as we have grown to understand the hard way, we can no longer throw things away entirely. If we can somehow convince everyone that the whole earth is "inside" our cultural space, the technology exists or can be developed for long-term ecologically sustainable waste disposal and prevention of pollution. Early hominids transported to the present would probably opt to blast our earth-generated garbage to the next higher "outside" space, actually outer space. In addition to just extending earth's pollution problems to nearby parts of our solar system, this "solution"

would eventually deplete the natural resources on earth that went
into producing the garbage in the first place.

The cultural part is the hard part. If the global "we"—ourselves
and the next several generations—can become more inclusive of
other cultures, traditions, and "races," then there is a hope that we
can forestall being overrun by mounting piles of garbage, poisoned
air and water, and increasing mortality from polluting agents. If not,
we will fall back into the ancient pattern of competition and replacement of populations characteristic of human evolution during
the Pleistocene. This is the all-too-familiar ethnocentric, winner-
take-all, rapacious adaptation of primitive culture. The former approach is more adaptive because world culture would gain from all
its components, ensuring a maximum supply of accumulated wisdom in adapting to the environment. One is tempted to say that this
is also the more intelligent approach, but intelligence is not the issue
here. Humans have been of the same intelligence since the end of
the Pleistocene, only 10,000 years ago, so one would be more accurate in commending a multicultural approach to solving world pollution problems on the basis of it being "more informed."

Endangered Species:
The Canaries of a Dying World

Coal miners used to take a caged canary down into the earth
with them to monitor the environment. If the oxygen became
depleted in the poorly ventilated mine shafts, the canary would
stop singing or die, since its higher metabolism required a higher
concentration of oxygen in the air. The miners then knew they
had only a short time to get out of the mine shaft or they themselves would die.

In the larger macrocosm of world environment, species of animals and plants represent coal-mine canaries. When they die it indicates that something is going wrong in our environment. The
only difference it that we do not have anywhere else to go.

The conservation movement has raised the preservation of species diversity to a goal unto itself. But the human species is really too selfish for that. Nonhuman species of animals or plants will never be preserved solely on the basis of their intrinsic right to exist, especially in the face of human needs. For example, the most ardent conservationist will hit a brick wall in arguing with a Montana sheep farmer who shoots an endangered wolf because it has just eaten one of his newborn lambs. And even the most pro-environment yuppie ranchers of central Oregon will call in the county animal control unit when a rare and endangered mountain lion carries off one of their prized llamas or emus. Conservationist groups attempt then to create a human character for their endangered species, to carve out a cultural identity. In order to popularize their efforts, they totemize their animals and make them part of human culture—baby seal plush toys are sent to contributors who want to stop the killing of fur seals, save-the-whale educators drive to schools in an old Volkswagen van painted like a whale, and virtually every group produces that ubiquitous identifying badge of cultural identity, a T-shirt, with a brightly screen-printed image of "their" endangered species on it. These are their totems, identifying them with a common cultural cause, conferring group membership, and giving their lives meaning.

There are some unspoken rules for which animal species get this sort of human attention. First of all, the species cannot be an ecological competitor. There are no foundations set up to ensure that the rat and the cockroach do not go extinct. Some species, like the mountain lion, wolf, and coyote, fall into an ambiguous category because they are ecological competitors only to a segment of the human population, ranchers in particular. Another rule is that the species to be conserved cannot be a human parasite. No one cares to prevent the extinction of the flea, the tick, or the liver fluke. A sine qua non of human conservation of a species then is lack of ecological competition.

It is not absolutely essential for its chances of preservation, but it helps if a species is more closely related to humans. Mammals, especially the intelligent, beautiful, and graceful ones like carnivores, primates, and sea mammals, get the lions' share of the conservation effort. Many birds, which can be beautiful and graceful if not intelligent, figure fairly prominently in conservation efforts. But there are twenty-two species of clams and thirty species of fish that are endangered in the United States, and despite the fact that their numbers reflect the cleanliness of human water resources—an essential human resource—campaigns to save them are few and far between. The endangered snail darter, a small fish living in the Tennessee River drainage system, is one exception. Its preservation in 1977 became the focal point of a major campaign to prevent the construction of a new dam. But in general nonmammals—reptiles, amphibians, fish, and invertebrates—are difficult species for people to identify with.

If scaly, slimy, and many-legged species of animals do not fare well with human conservation efforts, the immobile but animate world of plants is even more shunned. It is difficult to hug a threatened plant and have it gaze back at us appreciatively (as it is a baby tiger, but we can imagine it). Plants are also not cute, warm, and soft to touch. They just fall too far outside the range of our mammalian protective instincts. Only the majestic California redwoods have somehow transcended this plant–animal divide and command widespread reverence. But for most plants the approach to their conservation has to be on an intellectual and scientific plane.

Conservationists working to preserve tropical plants have found that the most effective approach is to aggregate a large number of species together into a unique and endangered habitat—the rain forest, for example. This way it can be shown that a large number of species growing together help to prevent such human ecological disasters as soil erosion or silting up of rivers and lakes. Appeals for the preservation of species diversity in the rain forest have also been based on the possible medicinal uses, again of potential human benefit, that may be found in single species of plants there. Again, plants will not be the subjects of conservation if they exist in a competitive ecological relationship with humans. No one has a particularly soft spot for poison ivy, for example. This and other species are culturally defined as "weeds" and are usually locally eradicated if possible.

Conservation of species diversity on earth is clearly of adaptive significance for human beings, simply because a diversity of species indicates an equilibrium of natural forces the same as those under which human evolution progressed to its present state. These conditions are most compatible with human health and well-being. There are also clear human uses for many wild species of organisms. As objects of study they provide knowledge of the evolution of life, instruct us in alternative ways of adapting to the environment, and can provide insights into environmental quality when used as an indicator species of pollution. Wild plants, in addition to their contribution to the makeup of their communities, can both serve as possible sources of future medicines and as possible future food plants.

In addition to the pragmatic reasons, some conservation of species closely related to humans is clearly more emotional. "Cute" species, such as the Chinese panda, receive an inordinate amount of attention, out of all proportion to their ecological significance. Evolutionarily, we can explain this behavior only as an extension of human feelings of caring toward intraspecific infants and juveniles, typified by the rounded features, soft contours, and short extremities that we consider "cute." Extending this same attitude toward less cute species will benefit us immensely in improving the eventual quality of our future environment.

HUMANS, THE INSATIABLE ENERGY EXTRACTORS

Energy extraction for humans is a tool for expanding the usable range of resources in their environment and to transform parts of the environment to their own purposes. When humans began using fire, a type of rapid oxidation that produces heat and light, they began to harness extracorporeal sources of energy. This control of energy provided the basis for humans to expand and adapt to the harsh environments of the Pleistocene. Fire was an important part of hunting (in driving game) and in the later development of agriculture, in that it was the major way that land was and is cleared in "slash-and-burn" agriculture.

Fire allowed food to be cooked. Previously inedibly tough foods, such as meat, and previously toxic foods, such as numerous alkaloid-containing plants, could now be ingested, broadening humans' dietary range. Fire gave humans a competitive interspecific edge over other animal species, it provided light at night, and it provided heat when humans moved north out of Africa. In more recent times, controlled use of fire in generating steam and then in the internal combustion engine ushered in the Industrial Revolution and the modern mechanistic age. Fire has served humankind well for over a million years, but there has been a price.

Because fire consumes oxygen and produces carbon dioxide and other chemicals resulting from the oxidation of its fuel, it contributes to changes in the atmosphere. From the beginning of humans' controlled use of fire in the early Pleistocene to only a few hundred years ago, the human contribution to fire's effect on the earth's atmosphere was infinitesimal compared to the effects of natural fires. Even today, forest fires and other naturally occurring fires are listed as major contributors to particulate and gaseous air pollution.

The major producers of today's air pollution, however, are of human origin. Automobile exhaust, emissions from industrial manufacturing, and smoke from power plants that burn fossil fuels are the three biggest contributors. Air pollution breaks down into the constituents of carbon monoxide, a poisonous gas; sulfur dioxide and nitrous dioxide, which produce corrosive sulfuric acid and nitric acid; particulate matter such as airborne carbon; and hydrocarbons from unburned fossil fuels capable of strong oxidation when subjected to reaction with ultraviolet radiation from the sun. Acid rain resulting from air pollution acts to kill acid-sensitive plants. Air pollution directly irritates our eyes and throats, and as already noted, can be fatal under some circimstances.

An indirect effect of air pollution has occupied much attention in the last decade, and that is the so-called "greenhouse effect." A greenhouse is typically a hot, high-humidity environment ideal for plants. Sunlight enters through the glass roof and walls but its radiant heat cannot leave. Carbon dioxide in the earth's atmosphere works much the same way as the glass in a greenhouse. It lets the sun through but then traps the radiant heat on the earth's surface. Since the Industrial Revolution the amount of carbon dioxide in the earth's atmosphere from burning of fossil fuels—coal, oil, and natural gas—has steadily risen. One estimate based on a study in Hawaii indicated that carbon dioxide content was increasing 0.02 percent per year.

In addition to higher average temperatures year-round, an increasing level of carbon dioxide would contribute to a substantial rise in sea level as increasing amounts of polar and glacier ice melt, and the precipitation regimes driven by temperature differences around the world, would change. If the greenhouse effect hit with full force, many coastal cities of the world and many trillions of dollars in personal property would soon be under water, and huge food-producing parts of the world could experience near-desert aridity.

The dire predictions of imminent global warming and the greenhouse effect are not universally accepted. Some climatologists question the accuracy of the temperature measurements, the slope of the extrapolation curve, and the assumptions of coupled climatic effects. But no one questions that eventually, at some time, a buildup of carbon dioxide from human energy consumption will bring about global warming. It is clearly ultimately going to be necessary to find and develop alternate sources of energy, ones that do not produce carbon dioxide, not to mention other polluting gases, as by-products.

Nuclear energy was a promising possibility in the 1960s and

1970s for making a shift from the age-old reliance on fire-based heat production for energy. But the dangers posed by nuclear accidents, radiation contamination, and long-term storage problems of nuclear waste have all but ended any further development of nuclear energy. Instead, solar energy in one form or another holds the most promise for the future of meeting human energy needs. Combined with the eventual need to disperse population out of cities and the resulting inefficiency of transporting energy from single sources by wires over a wide area, self-sufficient solar houses with vastly improved storage batteries would certainly be an expected future outcome. Solar-powered vehicles are long overdue and would effect a transformation in cities such as Los Angeles and Mexico City. In addition to avoiding the accumulation of greenhouse gases, increasing reliance on solar energy would wean the world from its reliance on petroleum, world reserves of which will eventually run out.

FUTURE HUMAN EVOLUTION: PREDICTING THE UNPREDICTABLE

There is an interesting paradox in popular views of current and future human evolution. One view is that change has become so rapid and so intrinsic to the human adaptation that almost anything is possible in the future. If one is an optimist, this can mean unbridled possibilities for technological progress and improvement of our quality of life. If one is a pessimist, then there will be almost infinite possibilities for human beings to destroy themselves and the parts of the planet on which they live, taking many other species with them. These opinions of rapid future human evolution are usually reserved for cultural evolution.

Another view of the future is that with modern medicine, natural selection has been slowed down to a standstill in terms of physical and biological evolution of the human species. Where once whole populations were decimated by infectious disease, today in many parts of the world antibiotics, public sanitation, and effective preventive medicine have reduced the mortality figures to almost nil. Parasitic diseases, the bane of human existence for untold millennia, have been virtually wiped out in the developed world as well. Trauma, which in the past was a major killer of human beings, is now effectively dealt with in the emergency rooms of hospitals or even en route by highly trained ambulance crews. Hemorrhaging can be quickly staunched, shock can be effectively treated, ingested poisons can be removed or neutralized, and body fluids restored. With these three major cullers of the human population—infectious disease, parasitic disease, and trauma—taken out of the picture, some argue that human biological evolution has largely stopped.

Both views are erroneous. Culture in fact is an ancient adaptation that changes its content but not its modus operandi from generation to generation. There will be a tendency for people to view their own ideas as right and correct far into the future. Ethnocentrism is a pan-human universal. We will either use our advanced brains to figure out that our ethnocentrism needs to extend to the entire human species, or one ethnocentric group will hold sway and eventually preempt the remaining global natural resources for itself, allowing the rest of the species to go to extinction. We can hope that the former path is chosen. But if human evolution and world history are any guides, the latter possibility is unfortunately just as likely.

Our vaunted view of modern medicine and its presumed powerful effect on human evolution is vastly overstated. It is true that people in the developed world no longer die of the same causes as people before the Neolithic Revolution. It is also true that average life expectancy is longer today than it was in prehistory. But these effects of modern medicine are extremely recent phenomena— less than two centuries old. They have also been in effect in a limited minority of the world's population. Even more importantly, however, modern civilization confers, along with all its advantages, several decidedly disadvantageous disease states. These diseases, such as cancer, heart disease, and diabetes, are killers. They can strike down the young as well as the elderly, and they are clearly a major force in natural selection. These diseases are virtually unknown among hunter–gatherers, and we may presume among our prehistoric ancestors as well. We may deduce then that these diseases are largely the result of interaction between our bodies and our environment—the environment that culture has created for us.

This interaction is complex, for medical geneticists have discovered a number of *oncogenes* that also seem to have a formative influence on many cancers. We must remember, however, that these same genes may have functioned effectively or may have never been "turned on" in our ancient environments. Natural selection is and will be clearly at work at the cellular and biochemical levels in effecting biological change in the human species. Medicine at best can forestall some of these effects but cannot cancel them out.

Physical evolution of the human species is predictive only insofar as we can forecast what will allow humans to more effectively bear and transmit culture, our primary adaptation to the environment now and in the future. Our brain, then, may well enlarge secondarily to selection for increased intelligence. The trite view of the humans of the future as large brained and puny bodied beings may indeed be as accurate as we can be. Certainly, intelligence is at a selective premium now as in the past, worldwide.

One thing is certain about the future. Humans will be at the mercy of the environment, as they and their ancestors have been since the dawn of time. Advanced culture evolved in the late Pleistocene to keep up with an accelerating pace of environmental change. But as human population and the spinoffs of human activities have increasingly changed our world, both our physical and cultural means of adapting to environmental change have been pushed to their limits. To restore a balance, the pendulum of cultural change will have to shift back to a pattern consistent with our evolutionary adaptations: low population density, a demographic pattern very different from that of today's maladaptive urban centers; environments that meet human needs, even at the massive costs of construction, environmental cleanup, and technological innovation needed for that result; and a shared cultural identity within the global human population.

These results are not prescriptions for an ecologically ailing world. They are the ultimate results of natural selection, assuming continued human survival on earth, the relative slow rate of biological evolutionary change, the finite natural resources of the earth, and the heavy toll of human death and suffering exacted by an increasingly poisonous environment. Humans can take remedial steps themselves by directing their own cultural evolution—the preferable way—or natural selection will eventually do it for them, either crafting future human evolution as it has over the past 5 million years, or simply snuffing out another fleeting species, as it has done innumerable times over the last 2 billion years.

Environmentalism needs to be pragmatic but at the same time appropriately humble before the vastly superior power of nature. Pragmatic because we cannot expect humans, themselves a product of natural selection, to risk their survival and reproduction for the benefit of other species. Humble because a nonanthropocentric, long-term view also unambiguously shows that our survival as a species is ecologically linked with that of other species, and it is the only view that explains why we are up against a number of Thomas Malthus's checks on population growth. Other species, if allowed to survive, will not only function as indicators of environmental quality but their unique adaptations will be of great potential practical benefits to human beings in the future.

Unfortunately, the short-term, arrogant view is still prevalent. Essayist Charles Krauthammer, in an article entitled "Saving Nature, But Only for Man,"* wrote, "Nature is our ward. It is not our master. It is to be respected and even cultivated. But it is

*Time, June 17, 1991.

man's world. And when man has to choose between his well-being and that of nature, nature will have to accommodate." Mr. Krauthammer argues for oil drilling at the expense of the caribou in Alaska and lumbering at the expense of the spotted owl in Oregon, but his precepts are flawed. *We* are the wards of nature, not the other way around. Nature is *our* master. Mr. Krauthammer may conceive of Mother Nature as a benign granny in her rocking chair who deserves our respect, but whose directives do not carry the force of law. Here he is sadly mistaken if he thinks that her flinty and wizened gaze would even blink at the disappearance of so inconsequential and short-lived an organism as "man." Old she may be but woe to the species that trifles with her world. Others have described her as "nature red in tooth and claw."

Surviving the environmental problems of our own making and living with our evolutionary baggage will be humankind's immediate challenge over the next several millennia. If world culture evolves into an effective adaptive mechanism and it stays in synchrony with our biology, then we might be up for our next big environmental challenge—the next glacial period, due some thirty centuries from now.

BIBLIOGRAPHY

Ayala, F. J. (1995). The myth of Eve: Molecular biology and human origins. *Science* 270:1930–36.

Bar Yosef, O. (1989). Upper Pleistocene cultural stratigraphy in Southwest Asia. In E. Trinkaus (ed.), *The Emergence of Modern Humans*. Cambridge: Cambridge University Press.

Bartholomew, G. A., and J. B. Birdsell (1953). Ecology and the protohominids. *Am. Anthropol.* 55:481–98.

Behrensmeyer, A. K., and A. Hill (eds.) (1980). *Fossils in the Making*. Chicago: University of Chicago Press.

Boaz, N. T. (1979). Early hominid population densities: New estimates. *Science* 206:592–95.

Boaz, N. T. (1985). Early hominid paleoecology in the Omo basin, Ethiopia. In Y. Coppens (ed.), *L'Environnement des Hominidés au Plio-Pléistocène* Paris: Fondation Singer–Polignac, pp. 283–312.

Boaz, N. T. (1988). Status of *Australopithecus afarensis*. In *Yearbook of Physical Anthropology* 31:85–113.

Boaz, N. T. (ed.) (1990). *Evolution of Environments and Hominidae in the African Western Rift Valley*. Virginia Museum of Natural History Memoir No. 1, 356p.

Boaz, N. T. (1992). Climate and evolution: First steps into the human dawn. *Earth* 1:36–43.

Boaz, N. T. (1993). *Quarry: Closing in on the Missing Link*. New York: Free Press/Macmillan.

Boaz, N. T. (1994). Significance of the Western Rift for hominid origins. In R. S. Corruccini and R. L. Ciochon (eds.), *Integrative Paths to the Past: Paleoanthropological Advances in Honor of F. C. Howell*. New York: Prentice Hall, pp. 321–43.

Boaz, N. T. (1996). Vertebrate paleontology and terrestrial paleoecology of As Sahabi and the Sirt Basin. In M. J. Salem, et al. (eds.), *The Geology of Sirt Basin,* vol. 1. Amsterdam: Elsevier, pp. 531–39.

Boaz, N. T. (1997). Calibration and extension of the record of Plio–Pleistocene Hominidae. In N. T. Boaz and L. Wolfe (eds.), *Biological Anthropology: The State of the Science,* 2d ed. Bend, OR: International Institute for Human Evolutionary Research, pp. 25–52.

Boaz, N. T., and A. J. Almquist (1997). *Biological Anthropology, a Synthetic Approach to Human Evolution.* Upper Saddle River, NJ: Prentice Hall, 608p.

Boaz, N. T., R. L. Bernor, A. S. Brooks, H. B. S. Cooke, J. de Heinzelin, R. Dechamps, E. Delson, A. W. Gentry, J. W. K. Harris, P. Meylan, P. P. Pavlakis, W. J. Sanders, K. M. Stewart, J. Verniers, P. G. Williamson, and A. J. Winkler (1992). A new evaluation of the significance of the Late Neogene Lusso Beds, Upper Semliki Valley, Zaire. *Journal of Human Evolution* 22:505–17.

Boaz, N. T., and L. H. Burckle (1984) Paleoclimatic framework for African hominid evolution. In J. C. Vogel (ed.), *Late Cainozoic Palaeoclimates of the Southern Hemisphere.* Rotterdam: Balkema, pp. 483–90.

Boaz, N. T., A. W. Gaziry, A. El-Arnauti, J. de Heinzelin, and D. D. Boaz (eds.) (1987). *Neogene Paleontology and Geology of Sahabi, Libya.* New York: Liss.

Boaz, N. T., D. Ninkovich, and M. Rossignol-Strick (1982). Paleoclimatic setting for *Homo sapiens neanderthalensis. Naturwissenschaften* 69:29–33.

Boaz, N. T., P. Pavlakis, M. McDonell, and J. Gatesy (1994). Early hominid environments in the late Pliocene of eastern Zaire: Analysis of the Senga 13B excavation. *National Geographic Research and Exploration* 10:124–27.

Brain C. K. (1981). The evolution of man in Africa: Was it a consequence of Cainozoic cooling? *Ann. Geol. Soc. S. Afr., Spec. Issue* 4:1–19.

Brain, C. K., and A. Sillen (1988). Evidence from the Swartkrans cave for the earliest use of fire. *Nature* 336:464–66.

Brown F. H. (1994). Development of Pliocene and Pleistocene chronology of the Turkana Basin. East Africa, and its relation to other sites. In R. S. Corrucini and R. L. Ciochon (eds.), *Integrative Paths to the Past.* Englewood Cliffs, NJ: Prentice Hall, pp. 285–312.

Brunet, M., et al. (1995). The first australopithecine 2500 kilometres west of the Rift Valley (Chad). *Nature* 378:273–75.

Cann R. L. (1993). Genes, dreams, and visions: The potential of molecular anthropology. In A. J. Almquist and A. Manyak (eds.), *Milestones in Human Evolution.* Prospect Heights, IL: Waveland Press.

Cartmill, M. (1994). Explaining primate origins. In *Research Frontiers in Anthropology.* Englewood Cliffs, NJ: Prentice Hall.

Cerling, T. E., J. Quade, S. H. Ambrose, and N. E. Sikes (1991). Fossil soils, grasses, and carbon isotopes from Fort Ternan, Kenya: Grassland or woodland? *Journal of Human Evolution* 21:295–306.

Curtis, G. H., and J. P. Evernden (1962). Age of basalt underlying Bed I, Olduvai. *Nature* 194:610–12.

Delson, E. (ed.) (1985). *Ancestors: The Hard Evidence.* New York: Liss.

de Menocal, P. B. (1995). Plio–Pleistocene African climate. *Science* 270:53–59.

Gabunia, L., and A. Vekua (1995). A Plio–Pleistocene hominid from Dmanisi, East Georgia, Caucasus. *Nature* 373:509–12.

Grine, F. (ed.) (1988). *Evolutionary History of the "Robust" Australopithecines.* New York: Aldine de Gruyter.

Grün, R., and C. B. Stringer (1991). Electron spin resonance dating and the evolution of modern humans. *Archaeometry* 33:153–99.

Hay, R. L. (1990). Olduvai Gorge: A Case History in the interpretation of hominid paleoenvironments in East Africa. In L. Laporte (ed.), *Establishment of a Geological Framework for Paleoanthropology.* Geology Society of America Special Paper 242, pp. 23–37.

Hill, A., and S. Ward (1988). Origin of the Hominidae: The record of African large hominoid evolution between 14 my and 4 my. *Yearbook of Physical Anthropology* 31:439–51.

Howell, F. C., and M. Edey (1965). *Early Man.* New York: Time-Life.

Huang, W., R. Ciochon, et al. (1995). Early *Homo* and associated artifacts from Asia. *Nature* 378:275–78.

Johanson, D. C., and M. Edey (1981). *Lucy: The Beginnings of Humankind.* New York: Simon and Schuster.

Johanson, D. C., and T. D. White (1979). Systematic reassessment of early African hominids. *Science* 203:321–30.

Kortlandt, A. (1972). *New Perspectives on Ape and Human Evolution.* Amsterdam: Stichting voor Psychobiologie.

Landau, M. (1991). *Narratives of Human Evolution.* New Haven, CT: Yale University Press.

Larick, R., and R. L. Ciochon (1996). The African emergence and early Asian dispersal of the genus *Homo. American Scientist* 84:538–51.

Leakey, M. G., et al. (1995). New four-million-year-old hominid species from Kanapoi and Allia Bay, Kenya. *Nature* 376:565–71.

Leakey, R. E. F., and R. Lewin (1977). *Origins.* New York: Dutton.

Ninkovich, D., and L. H. Burckle (1978). Absolute age of the base of the hominid-bearing beds in eastern Java. *Nature* 275:306–8.

Ninkovich, D., L. H. Burckle, and N. D. Opdyke (1982). Palaeogeographic and geologic setting for early man in Java. In R. A. Scrutton and M. Talwani (eds.), *The Ocean Floor.* New York: Wiley, pp. 221–27.

Potts, R. (1996). *Humanity's Descent: The Consequences of Ecological Instability.* New York: Morrow.

Radosevich, S. C., G. J. Retallack, and M. Taieb (1991). Reassessment of the paleoenvironment and preservation of hominid fossils from Hadar, Ethiopia. *American Journal of Physical Anthropology* 88:15–27.

Retallack, G. J. (1992). Comment of the paleoenvironment of *Kenyapithecus* at Fort Ternan. *Journal of Human Evolution* 23:363–69.

Rightmire, G. P. (1990). *The Evolution of* Homo erectus: *Comparative Anatomical Studies of an Extinct Human Species.* Cambridge: Cambridge University Press.

Ruff, C. (1993). Climatic adaptation and hominid evolution: The thermoregulatory imperative. *Evolutionary Anthropology* 2:53–59.

Shipman, P., and J. M. Harris (1988). Habitat preference and paleoecology of *Australopithecus boisei* in eastern Africa. In F. E. Grine (ed.), *Evolutionary History of the "Robust" Australopithecines.* New York: Aldine de Gruyter, pp. 343–81.

Sikes, N. E. (1994). Early habitat preferances in East Africa: Paleosol carbon isotopic evidence. *Journal of Human Evolution* 27:25–45.

Stringer, C. B. (1990). The emergence of modern humans. *Scientific American* 263:98–104.

Susman, R. L., J. T. Stern, and W. L. Jungers (1984). Arboreality and bipedality in the Hadar hominids. *Folia Primatol.* 43:113–56.

Swisher, C. C., G. H. Curtis, T. Jacob, A. G. Getty, A. Suprijo, and Widiasmoro (1994). Age of the earliest hominids in Java, Indonesia. *Science* 263:1118–21.

Trinkaus, E. (1981). Neandertal limb proportions and cold adaptation. In C. B. Stringer (ed.), *Aspects of Human Evolution.* London: Taylor and Francis, pp. 187–224.

Vrba, E. S. (1988). Late Pliocene climatic events and hominid evolution. In F. E. Grine (ed.), *Evolutionary History of the "Robust" Australopithecines.* New York: Aldine de Gruyter, pp. 405–26.

Vrba, E. S., G. H. Denton, L. H. Burckle, and T. Partridge (eds.) (1995). *Paleoclimate and Evolution with Emphasis on Human Origins.* New Haven, CT: Yale University Press.

Walker, A., and R. E. F. Leakey (eds.) (1993). *The Nariokotome* Homo erectus *Skeleton.* Cambridge, MA: Harvard University Press.

White, T. D., G. Suwa, and B. Asfaw (1994). *Australopithecus ramidus,* a new species of early hominid from Aramis, Ethiopia. *Nature* 371:306–12.

INDEX